Bicycle Accident Reconstruction for the Forensic Engineer

James M. Green PE; Janet Green RN

Order this book online at www.trafford.com
or email orders@trafford.com

Most Trafford titles are also available at major online book retailers.

© Copyright 2007 James M. Green PE; Janet Green RN.

All rights reserved. No part of this publication may be reproduced, stored in a retrieval system, or transmitted, in any form or by any means, electronic, mechanical, photocopying, recording, or otherwise, without the written prior permission of the author.

Fifth Edition.

Contributing Authors:
Robert Mionske, Esq.
Anand Kashbekar, Ph.D.
Kraig Willett
Janet Groves Green, RN, MSN, ANP-C, LNC

Print information available on the last page.

ISBN: 978-1-5536-9064-1 (sc)

Because of the dynamic nature of the Internet, any web addresses or links contained in this book may have changed since publication and may no longer be valid. The views expressed in this work are solely those of the author and do not necessarily reflect the views of the publisher, and the publisher hereby disclaims any responsibility for them.

Any people depicted in stock imagery provided by Getty Images are models, and
such images are being used for illustrative purposes only.
Certain stock imagery © Getty Images.

Trafford rev. 08/08/2018

 www.trafford.com

North America & international
toll-free: 1 888 232 4444 (USA & Canada)
fax: 812 355 4082

Table of Contents

List of Tables & Figures xv
Preface xxi

Chapter 1
Bicycle Accident Reconstruction: An Introductory Discussion 1
    Chapter Topics and Arrangement... 1
    Conclusion... 3

Chapter 2
The Role of Component Failure In Bicycle Accident Reconstruction 5
    Introduction... 5
    Types Of Component Failure In Bicycle Accident.................. 6
    Reconstruction... 6
    Conclusion... 16

Chapter 3
The Relationship Between Helmets and Bicycle Accident Reconstruction 17
    Introduction... 17
    The Cycle Helmet as a Factor in Bicycle Accident
    Reconstruction... 18
    Conspicuity... 18
    Ability of the Cycle Helmet to Withstand Impact................... 18
    Helmet vs. No Helmet in Bicycle Accident Reconstruction... 19
    The Bicycle Accident Reconstructionist and the Helmet
    Issue... 20
    Conclusion... 21

Chapter 4
The Impact of Roadway and Cycle Path Design on Bicycle Accident Reconstruction 23
    Introduction... 23
    Bicycle Path Design.. 23
    Bicycle Path Construction... 24
    Bicycle Path Maintenance... 24

Conclusion..25

Chapter 5
A Determination of Force Onto the Cycle-Rider or Pedestrian During the Impact of a Motor Vehicle 27
    Introduction..27
    Development of Force of Impact..27
    Conclusion...29

Chapter 6
Determination of the Reaction Times Available to a Cyclist at Different Intersection Configurations..31
    Introduction..31
    Analysis of Different Intersection Configurations...................31
    Conclusion...39

Chapter 7
A Determination of the Causal Factors of Bicycle Accidents Caused by Front Wheel Release and Measures to Prevent These Accidents 41
    Introduction..41
    Analysis of Falls Caused by Premature Release of the Quick Release Mechanism..41
    Preventing Premature Front Wheel Failure From Premature Quick Release...42
    Biokinetics of a Fall From Premature Release of a Quick Release Mechanism..43
    General Reconstruction of Accidents Due to Premature Release of Quick Release Mechanisms...44

Chapter 8
A Determination of the Casual Factors in Bicycle Accidents at Railroad Crossings 51
    Introduction..51
    Front Wheel Disengaging...51
    Improper Design or Maintenance of the Tracks......................52
    Weather Conditions or Cyclist Error Resulting in a Fall..........53

Chapter 9
The Utilization of the Principles of Conspicuity in Pedestrian and Bicycle Accident Reconstruction 55
    An Introductory Discussion..........55
    Properties of Light As Related to Bicycle and Pedestrian Accident Reconstruction..........55
    Factors Influencing The Visibility of Light..........57
    Blomberg studies..........58
    An Evaluation of Bicycle Retroreflector Systems and Their Effect on the Reconstruction of Bicycle Accidents..........60
    Using the Principles of Conspicuity to Reconstruct Bicycle Accidents..........62
    Principles of Conspicuity in Nighttime Bicycle Accident Reconstruction..........63
    Principles of Conspicuity in Daytime Bicycle Accident Reconstruction..........66
    Conclusion..........67

Chapter 10
Calibration of Nighttime Bicycle Accident Sites Using the Principles of Conspicuity and the Grid System 73
    Introduction..........73
    Conspicuity in Bicycle Accident Reconstruction..........73
    Reconstruction of a Night Accident: A Case Study..........75
    Engineering Analysis of the Facts of the Accident..........76
    Principles of Conspicuity in Bicycle and Pedestrian Accident Reconstruction..........77
    Conclusion..........78

Chapter 11
A Determination of the Reaction Times Available to a Motor Vehicle Driver Overtaking a Cyclist at Night 81
    Introduction..........81
    Conclusion..........83

Chapter 12
The Derivation of a Formula for Determining the Speed of a Bicycle Rider Down an Incline 85
    Nomenclature..........86
    Kinematic Terms..........86

Preliminary Calculation of the Mass Movement of Inertia of a
Wheel..87
Review of Force-Mass Acceleration Approach......................87
Work Energy Approach..90
A Field Study..91
Conclusion..92

Chapter 13
Basic Engineering and Physics Applicable to Bicycle Accident Reconstruction 95
Introduction..95
Equations..95
Shear Force on Forged-Rolled Steel and the Application of
These Forces to Bicycle Accident Reconstruction..............100
Conclusion..102

Chapter 14
A Determination of Braking Distance for Cyclists In Emergency Stopping Situations 103
Introduction..103
Deceleration Limits on Braking Distance...........................104
Coefficient Braking Formulas... 104
Stopping With the Rear Brake Only....................................106
A Summary of the Braking Distance Issue.........................108
Acknowledgements...108

Chapter 15
A Determination of the Structural Integrity of Bicycle Frames Subjected to Frontal Static Force 109
Introduction..109
Engineering Analysis..110
Application of Frontal Static Force.....................................110
The Sequence of the Frame Collapse During the Application
of Static Force..111
Results of Frontal Frame Collapse Testing.........................111
Conclusions From the Frontal Static Testing......................111

Chapter 16
The Engineering Dynamics of a Cyclist Being Thrown over the Front of a Cycle During a Sudden Stop 123
    Introduction..123
    Engineering Analysis..123
    The Kinetic Energy Method...125
    Fork Deflection Method...125
    The Straight Arm Method..126
    Method Comparison..128
    Kinematics of the Actual Fall...128
    Conclusion..129

Chapter 17
A Method for Determining Impact Damage to a Helmet When the Cyclist is Thrown Over the Front of the Cycle During a Sudden Stop 133
    Introduction..133
    Test Apparatus..133
    Determining the Approximate Impact Velocity for a Cyclist Pivoting About the Front Wheel... 134
    Method 1..135
    Method 2..135
    Conclusion..136

Chapter 18
A Determination of the Correct Method for Developing Motor Vehicle Speed of Impact in Bicycle Accident Reconstruction Collisions Using Engineering Literature and Field Verification..141
    Introduction..141
    Engineering Literature...141
    The Schmidt Method...141
    Example Calculation Using the Schmidt Method...................143
    The Searle Method..145
    Critique of the Searle Method..147

Chapter 19
Field Testing to Determine Throw Distance -The Haight and Eubanks Studies' 151
    Introduction..151

Conclusion..151

Chapter 20
A Determination of the Actual Point of Impact in Bicycle Accident Reconstruction Involving Motor Vehicles 159
    Introduction..159
    Analysis of Surface Points of Impact......................................159
    A Case Study Example of POIs Determination......................160
    Determination of the Actual Point of Impact.........................160
    Combination of the Velocity of the Truck and the Time of Crank Chain Ring's Fall...161
    Determination of the Actual Point of Impact Versus the Point of Impact From the Accident Report.......................................162
    Conclusion..163

Chapter 21
Determination of the Velocity of A Cyclist From Deformation of the Cycle From A Frontal Impact 165
    Introduction..165
    Derivation of Average Force Formula.....................................165
    Derivation of the Instantaneous Maximum Force Formula...167
    Conclusion..169

Chapter 22
A Determination of Bicycle Speed By Using Crush Data 171
    Introduction..171
    Facts of the Accident1..171
    Engineering Analysis..172
    Conclusions of Crush Data Testing and Determination of Pre-impact Cycle Speed..175
    Definition of Test Parameters..181
    Practical Achievement of Test (3)...181
    Equipment Specific to Accident Reconstitutions....................187

Chapter 23
The Civil Liability of the Forensic Engineer 189
    Introduction..189

Chapter 24
An Analysis of the Force Required to Release Cycling Shoes from Clipless Pedals **195**
    Introduction 195
    Engineering Analysis 196
    Analytical Results 196
    Conclusion 197

Chapter 25
The Use of EDSMAC to Simulate Bicycle And Motor Vehicle Collisions **203**
    Introduction 203
    Input Changes to EDSMAC 204
    Output of EDSMAC Trajectory Simulation 205
    Conclusion of the Analysis Using EDSMAC to Simulate Bicycle and Motor Vehicle Collisions 206

Chapter 26
The Role of Alcohol in Reconstructing Bicycle Accidents **213**
    Introduction 213
    Analysis of the Literature 213
    Factoring in Alcohol Consumption In Bicycle and Pedestrian Accident Reconstructions 214
    Case Study 215
    Summary 216

Chapter 27
An Evaluation of a Signed Release in Reconstructing Bicycle Accidents **219**
    Introduction 219
    The Importance of a Waiver in Reconstructing Bicycle Racing Accidents 221
    Reasonably Foreseen Hazards Covered by the Waiver 221
    Unforeseeable Hazards Not Covered by the Waiver 222
    Conclusion 222

Chapter 28
An Analysis of the Engineering Dynamics of a Bicycle Falling Sideways **223**
    Introduction 223

    Engineering Analysis..223
    Conclusion...225

Chapter 29
Metallurgy and the Reconstruction of Bicycle Accidents 229
    Introduction...229
    Frame Failures...229
    Exotic Materials...232
    Forensic Engineering Analysis of Frame and Fork Failure....232
    Metallurgy and Component Failure............................234
    Conclusion...234

Chapter 30
Determining Total Reaction Time for Accident Reconstruction and Civil Engineering Design 237
    Introduction...237
    Basic Human Perceive React Time.............................237
    Perceive React and Avoid Time at Night.....................238
    Conclusion...240
    Summary...241

Chapter 31
The Causal Factor of Bus Wheel Injuries And a Remedial Method for Prevention Of These Accidents 243
    Introduction...243
    Bernoulli's Principle..244
    Solving For P..244
    A Practical Design for Preventing Wheel Well Accidents.....246
    Conclusion...248

Chapter 32
Bike Fit 251
    Introduction...251
    Analysis of General Methods of Bicycle Fit................251
    Conclusion...254

Chapter 33
Bicycle Wheel Performance. 257

Chapter 34
Cyclist-Motor Vehicle Accidents 267
    Introduction..267
    Scope...267
    Motorists Cases..270
    Alabama...270
    Arizona...270
    Arkansas...271
    California...271
    Colorado...272
    Connecticut..274
    Delaware..275
    District of Columbia..275
    Florida..276
    Georgia...278
    Idaho..279
    Illinois..280
    Indiana...285
    Iowa...287
    Kansas..288
    Kentucky..289
    Louisiana..290
    Maine..296
    Maryland..297
    Massachusetts..297
    Michigan..298
    Minnesota..299
    Mississippi...300
    Missouri...301
    Montana...302
    Nebraska..302
    Nevada...304
    New Hampshire...304
    New Jersey...305
    New Mexico...305
    New York...306
    North Carolina...309
    North Dakota...311
    Ohio..312
    Oklahoma...315
    Oregon..315

Pennsylvania...316
Rhode Island...318
South Carolina..318
South Dakota..319
Tennessee...320
Texas..321
Utah..322
Vermont...323
Virginia..323
Washington..324
West Virginia...328
Wisconsin..328
Wyoming...330

Engineering Appendix A
Railroad Information 331

Engineering Appendix B
Appropriate Case Citations 341
Summary of Case Citations 343

Engineering Appendix C
Federal Regulations 379
Code of Federal Regulations 381
    Part 1512, Requirements for Bicycles............................381
    Subpart A-Regulations..382
    Consumer Product Safety Commission...........................390
    Subpart B-Policies and Interpretations...........................390

Appendix D
Engineering Bibliography 393

About the Authors 403

INDEX 405

List of Tables & Figures

Chapter 2
    Figure 1 - The parts of a typical bicycle.....................6
    Figure 2 - Checking for frame alignment....................8
    Figure 3 - The effect of an 800N rotating force.................9
    Figure 4.1 & 4.2 - Pictures of automated spoking machines...10
    Figure 5 - Wheel truing................................10
    Table 1 - Spoke tension determined by displacement............11
    Figure 6 - Parts of the derailleur system........................12
    Figure 7 - Chain routing at derailleur..........................14
    Figure 8 - Chain deflection..................................15

Chapter 6
    Table 1 - A definition of the reaction times and time of travel of a class 4 vehicle and a cyclist at different intersection configurations..........................33
    Figure 1 - Two lane, four way intersection.................34
    Figure 2 - Determination of reaction time - 4 lane intersection 35
    Figure 3 - Four lane turn into 2 lane........................36
    Figure 4 - Determination of reaction time - T intersection........37
    Figure 5 - Determination of reaction time from 2 lanes to 4.....38

Chapter 7
    Figure 1 - Principle of quick-release........................41
    Figure 2 - Front fork positive retention device - Brilando patent..46-48
    Figure 3 - The design of a closed end front fork assembly 49

Chapter 8
    Figure 1 - Wave length and frequency as a function of color...69
    Figure 2 - An illustration of the test track utilized for conducting the retroreflector and bike lamp tests........................70
    Table 1 - Photometric test results CAT eye reflector................71
    Table 2 - A comparison of the CAT eye retroreflector with bicycle headlamps at various distances.....................72

Chapter 12
Figure 1 - Velocity of a 150 pound rider at various slopes coasting from a stop for a distance of 100 feet.........94

Chapter 13
Figure 1 - Spatial relationship on the trajectory of a cyclist.....97
Table 1 - Degree of slope as a function of percent of slope.....98
Table 2 - Shear v. the type of material....................................101

Chapter 14
Figure 1 - Rim brake types...103

Chapter 15
Figure 1 - Checking for downtube damage.............................109
Figure 2 - Bike frame loaded statically and dynamically........110
Figure 1 - Overview of frame jig used for static frame tests...113
Figure 2 - Testing and recording equipment used in the front load static test..113
Figure 3 - Bicycle frame test results..114
Figure 4 - Bicycle frame test results..114
Figure 5 - Individual collapsed frames............................115-120

Chapter 16
Figure 1 - Example of a rider beginning forward rotation......124
Figure 1 - Location of critical points for determining the engineering dynamics of a front fall.................130-131

Chapter 17
Figure 1 - Pendulum type testing apparatus............................138
Figure 2 - Testing apparatus after release of head form.........139
Figure 3 - Head form and test helmet as pad is moved under helmet..139
Figure 4 - Hydraulic ram in lower right corner pushes pad to simulate sliding...140
Figure 5 - Head form secured to adjustable neck joint...........140

Chapter 18
Figure 1 - Pedestrian and cyclist trajectory components by Schmidt...143

Figure 2 - Determination of center of mass of the cycle rider system..145
Figure 3 - Coefficient of friction v. critical projection angle..147
Table 1 - Cyclist, pedestrian - trajectory parameter range using the Schmidt method..149

Chapter 19
Figure 1 - Motor vehicle approaches cycle............................152
Figure 2 - Initial impact with cyclist being vaulted onto the hood..152
Figure 3 - As the motor vehicle stops, the cyclist is ejected from the hood and windshield..153
Figure 4 - The cyclist continues the ejection trajectory onto the ground..153
Figure 5 - Final resting place of the cyclist and cycle............154
Figure 6 - Crush data and windshield damage to subject motor vehicle..154
Table 1 - A determination of throw distance of a cycle and cyclist at various speeds.....................................155-157

Chapter 20
Figure 1 - Relationship between the accident report POI and the actual POI..162

Chapter 22
Figure 1 - Bicycle test track...177
Figure 2 - Speed vs. fork deflection.......................................178
Figure 3 - Velocity vs. fork deflection for geneic frames.......179
Table 1 - Speed vs. fork deflection using generic frames.......180
Addendum - Methodology of reconstitution of actual accidents on test tracks..181-188

Chapter 24
Figure 1 - Clipless pedal test jig..198
Figure 2 - Overview of testing jig...199
Figure 3 - Side view of testing jig...199
Figure 4 - A comparison of the lateral release values for various 1993 clipless pedals......................................200

Table 1 - A comparison of the lateral release values for various
1993 clipless pedals..201-202

Chapter 25
Table 1 - Predicted time of travel from EDSMAC to simulate
bicycle and motor collisons.....................................206
Figure 1 - Predicted time of travel from EDSMAC compared to
field measured values..208
Figure 2 - Example trajectory...209

Chapter 26
Table 1 - Alcohol distribution ratios in various organs and body
fluids after equilibrium..215

Chapter 28
Figure 1 - Sum of the forces method.....................................227
Figure 2 - Center of gravity method......................................228

Chapter 29
Table 1 - A comparison of strength vs. weight of materials used
in bicycle frames...230
Table 2 - Yield and tensile strength for steel frame tubing....231

Chapter 30
Table 1 - Expected response distances (FT) to pedestrian
targets...239

Chapter 31
Figure 1 - Velocity v. inverse pressure..................................245
Figure 2 - Bus wheel well with S-1 Gard..............................247

Chapter 32
Figure 1 - Component Identification for Bike Fit..................255

Chapter 33
Table 1 - Wheel Data..259
Table 2 - Variables Affect Overall Performance...................260
Appendix A - Equation of Motion...262

Appendix B - Solo Training Ride Data...................................263
Appendix C - Uphill Portion of Training Ride........................264
Appendix D - Criterium Data..265
Reference Case Data...266

Preface

This book is part of an ongoing effort to apply Professional and Forensic Engineering analysis to the science of bicycle accident reconstruction. In order to insure the accuracy of the engineering information provided, a review process was set up using engineers from The American Society of Civil Engineers and The National Academy of Forensic Engineers. I am particularly grateful for the peer review conducted by Anand Kasbekar, Ph.D., Joel Hicks, PE., and, in the area of engineering ethics, Marvin Specter, P.E., L.S. I also appreciate the many comments provided by Engineers across the country and the Taiwan Bicycle Institute.

The previous edition of this text included a legal analysis of the issues in bicycle acident reconstruction. In order to update the legal analysis, Robert Mionski, Esq., has authored a chapter on the legal issues of bicycle accidents. Bob was a National Cycling Champion and a member of the United States Olympic Cycling Team. To state that he is well qualified to address the legal issues of reconstructing bicycle accidents is an understatement.

As I stated in previous editions, this effort is ongoing. Reconstructing bicycle accidents is not an exact science due to many variables involved in the analysis. Constructive input to this continual effort is encouraged.

The metallurgy chapter has been rewritten to reflect the changes brought about by the new composite frame materials.

The issue of conspicuity is still a significant concern in the reconstruction of pedestrian, bicycle and motorcycle accidents. The conspicuity issues and their application in reconstructing these accidents are still valid in this edition.

Since the publication of the 4th edition on this subject, the principles defined in that edition have been subjected to use in reconstructing thousands of accidents by myself and other Professional Forensic Engineers. I have attempted to apply input to this field as I have received constructive criticism from Engineers, literally, from around the world. Most of the principles expounded in the earlier edition have held up very well in the real world. It is expected that this edition will change as I receive constructive input in the years ahead.

Chapter 1

Bicycle Accident Reconstruction: An Introductory Discussion

Since there are many fine texts available on the general subject of accident reconstruction, it is not the purpose of this book to define basic forensic engineering techniques. My intention is to develop a reference text for Professional Engineers to assist them in understanding how to reconstruct bicycle accidents. (Due to the close parallel in engineering dynamics, a large part of this text can also be applied to pedestrian and motorcycle accident reconstruction.)

The engineering dynamics of bicycle movement are entirely different from other types of phenomena. The Reconstructionist sometimes incorrectly applies the same engineering forces to a bicycle that he or she would apply to a car in reconstructing accidents. However, unlike a car, a cycle is essentially a frictionless entity with the mass of the rider being the dictating factor in any engineering equations. As a result, most dynamic equations used with motor vehicles simply do not apply.

Chapter Topics and Arrangement

I have organized this book into chapters that cover specific issues.

Another area that is emerging as a critical issue in the reconstruction of cycle accidents is the conspicuity (the ability to see and be seen) at the accident site Engineers have worked, literally for decades, to define safe and effective coloring and lighting systems that enable the pedestrian and cyclist to be conspicuous under both day and night conditions. There is a chapter summarizing this work as well as a chapter detailing a peer-reviewed method for accurately and consistently calibrating an accident site where conspicuity is involved.

Much has been made of the importance of a cyclist wearing a helmet, not only in terms of preventing injury, but also in defining contributory negligence. While the capability of a helmet to prevent injury is still being researched by the engineering community, the chapter on the use of cycle helmets is an initial step in determining the impact that a bicycle helmet may (or may not) have on a bicycle accident.

If there is a particular type of cycling accident that is of epidemic proportions in this country, it is the type involving premature release of the front bicycle wheel. My chapter on this subject details the problem and suggests remedial measures to prevent these accidents.

Several areas of bicycle accident reconstruction center around roadway design, reaction times, intersection designs (including railroad crossings) and force-of-impact. One chapter offers several basic engineering equations related to these subjects. Since the equations for reconstructing bicycle and pedestrian accidents are entirely different from those employed in any other forensic discipline, I have asked for peer-review from other professional engineers on the engineering dynamics, wherever possible. In addition, I have used a tremendous amount of information from the academic community (in the form of operations research modeling) which I have verified in the field using actual cycling accidents.

Many of the equations listed in the Chapter 13 (Basic Engineering and Physics Applicable to Bicycle Accident Reconstruction) are unique to pedestrian and cycle reconstructions. Field verification has been done and I have found these equations to be accurate. In fact, force-of-impact onto the human body has been accurate in back-calculating to determine the speed of the impact. (This concept has been laboratory tested very successfully using cadavers.)

The equations in the previous edition of this book were subjected to considerable scrutiny in recent accident reconstructions. As a result, the equations in this edition reflect a great deal of additional field testing, as well as input from a number of highly reputable professionals. I fully expect the work to evolve further as I continue to collect and evaluate data from other professionals in this field.

The issue in one important case, *Johnson vs. Derby Cycle Corp.*, was whether or not a cycle company should force a consumer to purchase a bike light with his/her bicycle. The jury answered this question in the affirmative and awarded the plaintiff approximately $7 million dollars.

The *Johnson* verdict suggests that cycle companies may be held accountable for a consumer's failure to use a bike light at night. The verdict has, thus, greatly impacted bicycle accident reconstructions involving conspicuity issues.

The testing on retroreflector systems which was conducted in connection with the *Johnson* case is included in this edition. Also included is a discussion of the visibility of these systems both in and outside of the beams from a motor vehicle's headlights.

I also prepared, in connection with the *Johnson* case, an EDSMAC simulation of the subject motor vehicle/bicycle collision. The EDSMAC program is widely used by accident reconstructionists and accepted by many courts. The modifications I have made to the program for use in motor vehicle/bicycle accident reconstructions is also discussed in this edition. The accuracy of the motor vehicle/bicycle collision simulation is excellent pre-impact. Additional work is needed post-impact.

In the *Johnson* case, actual bicycle crush testing was conducted on a test track. The data from these tests was used to determine the speed of the subject bicycle prior to impact. A discussion of this work is also included.

The above summary indicates what is already my firm belief-that the work in this edition represents the "state of the art" for the engineering, scientific, and legal professions in the area of bicycle accident reconstruction.

Conclusion

I feel that this book will be useful for Forensic Engineers who need to handle bicycle accident cases. It also should aid Attorneys in deciding the advisability of moving forward with a lawsuit. Very often, when I am asked to reconstruct bicycle accidents, I find defendants have been named who are not at fault. One of my goals with this text is to lessen this problem through the application of the principles discussed here.

In addition to assisting the above-named professionals, it is my hope that the information contained in this book will be used by the cycling industry to make the sport much safer. Manufacturers of cycling related equipment do not always heed literature on conspicuity and visibility which, if applied at little or no additional cost, could make cycling a much safer sport.

In another area of safety, there are entirely too many tragic accidents that occur due to component failure on cycles. I believe that the cycling industry should initiate basic quality assurance procedures in the manufacture of these components. Such procedures are standard in other industries (i.e., automobile plants and other forms of manufacturing) but are noticeably lacking in bicycle manufacturing.

In the reconstruction of these bicycle accidents, I have observed that some bicycle companies do make an effort to comply with known standards and conspicuity values. Schwinn Bicycle Co., Raleigh Cycle Corporation, and Cannondale will change product design and revamp quality assurance procedures when problems, resulting in cycling accidents, are brought to the attention of management. For instance, Schwinn's Paramount line of helmets carry conspicuous colors. In addition, Schwinn has installed positive retention devices on quick-release mechanisms to prevent premature front wheel release. This has all but eliminated the problem for Schwinn.

Raleigh Cycle Corporation and Cannondale Corporation have also instituted similar socially responsive procedures. They have instituted stringent quality control procedures to insure positive retention devices are installed on front quick release mechanisms. Their quality control procedures also insure that all componentry from their suppliers meet or exceed industry standards. Laboratory testing of their products reveals the structural integrity to be among the highest in the industry. The leading component manufacturer, in terms of quality assurance, is best exemplified by SRAM Corporation in Chicago, Illinois. The quality assurance program and component testing enables this company to predict accurate operation ranges on all their products.

Please note that any opinions and conclusions given in a chapter are solely those of the chapter's author. Such opinions have not been reviewed or endorsed by the book's other contributors. I have acted as an editor in reviewing and checking the technical accuracy of the contributing authors.

Chapter 2

The Role of Component Failure In Bicycle Accident Reconstruction

Introduction

The incidence of component failure in bicycle accident reconstruction is relatively small compared to the total number of accidents. Thus, detailed analysis of this type of accident is, unfortunately, lacking in the Engineering and accident reconstruction literature. However, an emergency room analysis of different types of accidents has been done.[1] The results of this analysis show that, based on the data reviewed, bicycle accidents caused by component failure comprise approximately 17% of total emergency room cases. Analysis of the component failure accidents that I have personally investigated reveal that this rate of occurrence is accurate. It should be noted that component failure is often attributed to an accident when the actual cause of the accident is inadequate equipment maintenance by the rider.

The purpose of this discussion is to define the types of component failure that directly contribute to bicycle accidents. It should be noted that most of the failures are experienced by bicycle racers when the equipment is stressed during racing.

It is recommended that the Forensic Engineer use the services of a certified and competent bicycle mechanic to identify component failure. Proficiency in bicycle mechanics comes from years of "hands on" training under the direction of experienced bicycle mechanics. The bike, although a simple machine at first view, is in fact a finely tuned apparatus with very exacting tolerances. There are available several authoritative books and manuals on servicing and building bicycles. These books provide the appropriate tolerances and fit for the various component parts.[2] However, the best source for the fit of cycle component parts is still the professionally trained mechanic. Competent mechanics on the professional racing teams are given a very exalted standing due to their capabilities.

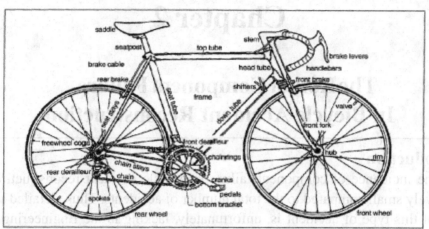

Figure 1 The parts of a typical bicycle, showing their most common designations. Although only one specific model is shown here, other bikes have largely the same parts. (Illustration by Rob Van der Plas)

Types Of Component Failure In Bicycle Accident Reconstruction

The following classes of component failure are drawn from accidents that I have personally investigated. Please note that a metallurgist should be retained when component failure is an issue in an accident reconstruction.

Frame Failure

With the exception of frame failure due to impact from a motor vehicle, bicycle frames tend to fail at the points which receive the greatest stress during riding. Models of these stress forces have been generated by various researchers. [3] These forces can create fatigue points at all the lug joints. The lug joints are those points on the frame where the tubes and the joints are brazed, soldered, or glued together.

Since the transmittal of stress forces tends to impact these areas of brazing or gluing, frame builders have traditionally overbuilt these areas. If the brazing or gluing process is not done correctly, the tubes can literally pull out of the lugs during normal riding. In an effort to strengthen these fatigue points, a process was recently developed which employs aluminum tubes glued to metal lugs. The first company to employ this technology was the Vitus frame building company in France. To identify whether the fatigue points on the bike are at risk of failure, the frame tubing should be inspected where it is installed in the lug for any signs of either brazing material or glue that is not completely around the tubing. The tubing can then pull out of the lugs when stressed.

Frames that I have tested will take forces lateral to the plane of the frame in excess of 5000 lbs. at the lug section without failing. The gluing and brazing methods of tube and lug attachment are entirely adequate for the forces transmitted during riding. It is when the gluing or brazing is not done properly that failure, often catastrophic, can occur.

Other types of frame failure can occur from misalignment of the front dropouts to the chain stays. The alignment of these dropouts must be perfect in order for the cycle to steer straight or turn under control through curves. If improper steering is alleged to be a problem, the alignment of the front dropouts should be checked by the reconstructionist using a Campagnolo set of alignment tools (or equivalent).

Some types of frame failure have been noted with the lugless one-piece frames. Since approximately 1987, some frame manufacturers have been making thin-walled one-piece aluminum and composite material frames. These frames are manufactured by casting the material in a mold under varying manufacturing conditions. The frames are as structurally solid as the metal lug frames, with one exception. If the frames are placed in a bicycle repair stand with a frame gripping device, they will split or break at the repair stand clamp. The Forensic Engineer should investigate the history of this type of frame when a failure occurs. In the frames that I have evaluated, there were no warning labels that instructed the user about this potential problem.

Another type of frame failure is the total misalignment of the frame. This occurs when the tolerances are not correct in the casting of a molded frame, or the brazing and gluing of the lug frame. Simply put, the frame is different on one side than it is on the other (it is asymmetrical). When this occurs, the rider can experience vibration, wobble, or continual loss of control while riding.

The primary reason for this type of frame misalignment is the frame builder's failure to utilize the correct tube measurements. The Forensic Engineer can determine the existence of this problem by using the Campagnolo frame alignment tool (or equivalent). Of course, if the Forensic Engineer investigates the history of the frame, he may find that it has been involved in a wreck. This may also cause total frame misalignment.

Investigation of those accidents that result from chronic loss of steering or handling characteristics should center, in part, on the potential of frame alignment problems. The importance of a rider "holding the line" (the ability to ride straight in a large pack of riders) has been investigated by mechanics on the professional race circuit. As a result, the proper tools and assistance of a bicycle mechanic, or metallurgist, can help the reconstructionist decide if frame misalignment or failure is a causal factor of the accident (See Figure 2).

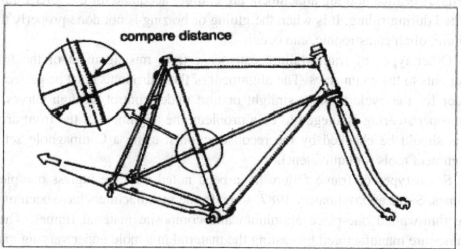

Figure 2 Checking for frame alignment. Ideally, the dimension should be identical on both sides. Any discrepancy of more than 3 mm (1/8 in) is definitely unacceptable and should be corrected. (Illustration by Rob Van der Plas)

Wheel Component Failure

Catastrophic failure of front wheel components is the most serious problem that can impact a cyclist while riding. The failure of front wheel components can cause a cyclist to be thrown over the front of the bike and sustain very debilitating upper body injuries. Bicycle racers will push away, or block with their arms, other riders from the front wheel in an attempt to prevent this type of injury.

A common type of front wheel component failure is the collapse of the front wheel. When wire spoke wheels are built, it is necessary to prestress and re-true the wheel to prevent its collapse from lateral forces. These forces are component vectors that are applied perpendicular to the plane of the wheel during normal riding motion. Although the primary force applied to the wheel is the weight of the rider, the side sway of the cycle during pedaling and turning causes lateral forces to be applied. Although some differences have been observed from modeling and measuring this phenomena,[4] the lateral force is generally 1/30th of the rider weight force. Although this lateral force is a relatively small vector in comparison to the overall weight on the wheel, it can cause complete wheel collapse if the wheel has not been properly constructed (See Figure 3). When reconstructing potential wheel collapse cases, the Engineer should investigate the history of the cycle's wheels. If the remaining wheel was built at the same time as the collapsed wheel, it can tell the investigator the condition of the collapsed wheel. When wheels are manufactured in lots, the spokes are laced onto the wheel using a lacing machine. These lacing machines are excellent for placing spokes into predetermined lacing patterns

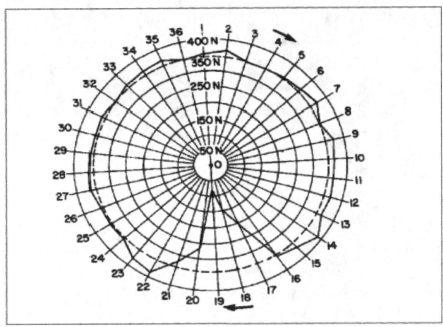

Figure 3 *The effect of an 800 N rotating force on the tension of individual spokes when these are prestressed at 350N. (Illustration by Rob Van der Plas)*

on the wheel. They also leave each spoke at a predetermined tension once the wheel exits the lacing machine (See Figure 4). However, these machines fail to laterally stress and re-true the wheel several times prior to its use. In order for a spoked bicycle wheel to operate adequately without collapse, the wheel must be laterally stressed and re-trued during its manufacture. This is an activity that is best done by a competent wheel builder. A wheel properly stressed and retrued will withstand several years of normal riding with occasional re-trueing. Also, a wheel that is properly stressed and retrued during construction will not have disparate tension values in its spokes after riding. A wheel that is subject to collapse in the future will have tension values that vary significantly from spoke to spoke. Again, examination of the wheel that has not collapsed will reveal the probable condition of the collapsed wheel. This assumes that both wheels were built at the same time, in the same manner, and have been subjected to the same wear and tear (See Figure 5).

Various measuring devices are available for determining the tension values of spokes. I prefer the Hozan tension meter, because it is extremely accurate with proper calibration. Measurement of the spoke displacement values can be in psi or mm. Some acceptable values for spoke tension are as follows:

Figure 4.1 and 4.2 All major manufacturers use automated spoking machines these days. Despite their accuracy, some manual truing and final tension is often required. (Photos by Rob Van der Plas)

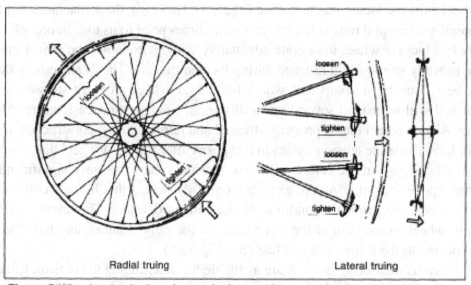

Figure 5 Wheel truing is done by tensioning and loosening spokes selectively, as shown here for radial and lateral deflections, respectively. (Illustration by Rob Van der Plas)

Table 1
Spoke Tension Determined by Displacement

Wheel Section	Displacement (mm)
Left Front	0.80
Right Front	0.80
Right Rear	1.25
Left Rear	0.80

The high tension value on the right rear wheel is due to the need for a stronger, stiffer right side. This is the side on which the cluster is located. The cluster contains the circular gears that transmit force to the wheel causing a driving force to the cycle. Typical values of a wheel about to collapse will vary from 0 to greater than 1.3 mm of displacement. While the values given in Table I are not meant to be final industry standards, the displacement values should be consistent on each side of the wheel.

It is not acceptable for a bicycle wheel to collapse without warning during normal riding. Properly built wheels that are stressed and trued will travel through several years of riding before problems result from normal use.[5]

Front Fender Failure

A catastrophic bicycle accident may be caused by the binding (collapse) of a front fender against the front fork. The fender or brace may bind into the front fork by catching on the tire.

During normal riding, vibration occurs on the bike front fender. If knobby mountain bike tires are used on the front of the cycle, the fender on the brace can catch on the tire's protuberances. When this happens, the fender and brace collapse into the front forks, the cycle comes to an abrupt stop, and, as a result, the rider is thrown over the front of the bike.

Testing the forces involved in moving a brace and front fender into the wheel reveals that properly configured fender braces directly impact the amount of force necessary to move the assembly into the wheel. If the brace is a single blade of metal with a high degree of flexibility, then the blade of the fender could be caught on the knobby tire causing catastrophic failure. The best design for this type of assembly is the round brace with little side-to-side movement. This brace will not catch on the knobby mountain bike tire.

The Forensic Engineer should carefully evaluate the history of the cycle and the structural integrity of the member brace for flexible movement at different force values. Inflexible rounded braces will not catch on knobby tires;

flat bladed fender braces with a high degree of flexibility do have the potential to catch on the tire.

It is debatable how effective and necessary the front fender on street and off-road cycles is for preventing road spray and grit from impacting the rider. The "rooster tail" (road spray) during inclement weather typically passes into the bottom part of the cycle. Consequently, the front fender is not really necessary. In light of this fact and the considerable risk of failure, the need for a front fender is dubious, indeed.

Seat Failure

Accidents also occur from failure of the seat post at the point of connection to the seat. In these instances, the rider is typically impaled on the seat post. Seat collapse does not generally occur when the seat is fastened to the seat post with two parallel bars that hold the seat away from the seat post. In comparison, seat collapse often occurs during normal use when the seat post is fastened directly to the seat. Thus, an inspection of the seat post configuration is usually sufficient to determine the causal factor of this type of accident. The inspection should, however, be coupled with a metallurgist's report on the structural integrity of the bike.

Front Derailleur Failure

The mechanism located on the seat tube directly above the two large chain rings is responsible for changing the chain from the large to the small chain ring. The derailleur is also used to move the chain from the small to the large chain ring (See Figure 6). Accidents occur when the derailleur is incorrectly adjusted such that the chain continues past the cogs on the chain ring and fails to catch on

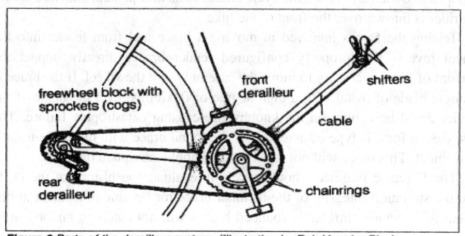

Figure 6 Parts of the derailleur system. (Illustration by Rob Van der Plas)

the cogs on the ring. When this occurs, the rider is suddenly pitched forward since the pedals no longer encounter resistance. Serious accidents usually result from this imbalance in the rider cyclist system. More experienced cyclists can physiologically adjust to this imbalanced situation by quickly shifting rider weight when the chain jumps off the chain ring. It should be noted that this type of weight adjustment requires the abilities of a well-trained athlete. Most untrained cyclists are not aware of the need to shift weight off the pedals while shifting from one chain ring to the other. This weight shifting is necessary to go from one chain ring to the other, even with a properly adjusted front derailleur.

The reconstructionist should investigate the adjustment of the front derailleur when attempting to define the reason for the chain skipping or jumping off the front cogs. If, after testing, the chain is found to jump due to improper adjustment, the history of this problem should be defined. If the rider is knowledgeable about the problem, then he or she should bear some responsibility for the problem. If the chain jumps on a new cycle and the rider is not properly trained to make the necessary adjustments, then responsibility for the problem should be assigned to the instructions in the owners manual, or the original assembler of the cycle.

Rear Derailleur Failure

The mechanism that causes the chain to engage different gears is located on the frame immediately next to the rear wheel. This mechanism essentially moves the chain to different cogs by means of the rider applying force to a shifter that is usually located on the right side down tube (See Figure 7).

There are two primary types of shifters on the market which serve purpose. These are the fiction shifter, made popular by Campagnolo the above and other European manufacturers, and the index shifter, introduced on a mass level by the Japanese manufacturer, Shimano. The friction shifter relies on the force between two teflon or plastic discs in the braze on shifters in the down tube. These discs will hold the shift lever in place if the friction is adjusted properly. Under these conditions, the rear derailleur is also held in place during riding. Problems result when the discs wear out or the friction adjustment knob is not adjusted properly. When this occurs, the chain can skip on the rear cluster, thereby causing the rider to lose balance and potentially fall. The best way for the reconstructionist to identify this type of failure is to either ride the cycle while vigorously shifting the gearing sequence or put the bike on a simulator and reproduce the shifting sequence.

The index shifter, pioneered by Shimano, uses a serrated disc with a spring mechanism that allows each gear shift to be made by shifting the spring into the individual indents on the serrated disc. An audible click can be heard each time the shift is made. This type of mechanism offers several advantages, not

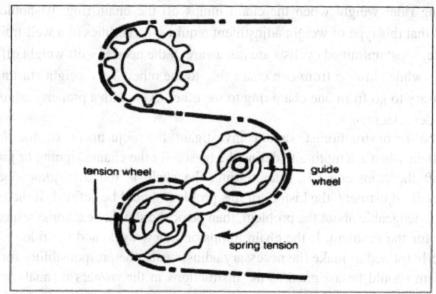

Figure 7 Chain routing at derailleur. (Illustration by Rob Van der Plas)

the least of which is the ability of the shifting mechanism to stay in position due to the indentation. It is very difficult for a chain to jump off a cog in the rear cluster with this type of configuration unless a misadjustment of the shift levers has been made during assembly. Also, the shift cable going back to the rear derailleur can stretch with use and cause the chain to jump. The various manufacturers of the index shifter systems use an adjustment mechanism on the cable entering the rear derailleur system to allow for tightening of the cable with use. As with the friction derailleur system, the best way to determine the causal factors of this type of accident is to change gears, either while using a simulator or while riding the cycle.

When an inexperienced and nonathletic person shifts weight during the normal pedaling motion, and the derailleur system does not function properly, the cyclist usually becomes seriously unbalanced, or falls. I have interviewed many accident victims who did not understand how a derailleur system worked. Therefore, a cyclist should be shown how to properly use and adjust derailleur systems. This can be accomplished by instruction from the bike shop and reading the owner's manual (See Figure 8).

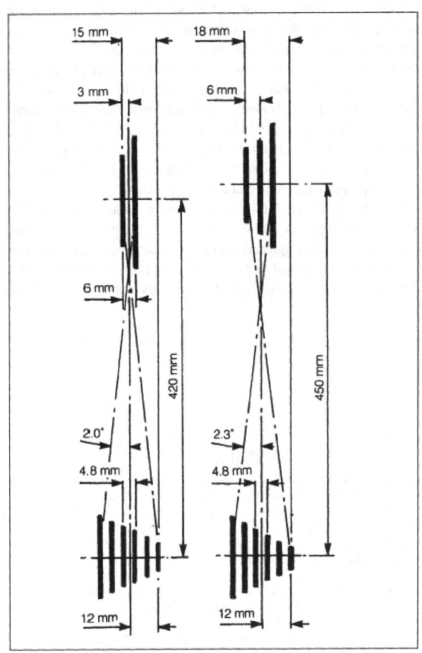

Figure 8 Chain deflection as a determining factor for the suitability of gearing combinations. (Illustration by Rob Van der Plas)

Conclusion

Eventually, all cycling components will wear out during normal use and will need to be replaced. The reconstructionist will need to be aware of such foreseeable component failure. The reconstructionist must also consider possible unforeseeable causes for component failure. For example , it is not foreseeable on a new cycle will collapse without warning after riding the bike that wheels only a few yards. Additionally, it is not foreseeable that a frame will fail or a cycle will be incapable of being steered along a straight line. It should be noted that most component failures that I have investigated have occurred on "low end" cycles that were either sold right out of the box at a discount store or were improperly assembled at a store or bike shop.

As previously noted, the Engineering of the bicycle is entirely different from any other type of product. In no other place is the physiology of the user so important in the actual design and operation of the product as it is in cycling. Thus, in determining the causal factors of accidents involving component failures, the reconstructionist should employ the services of a skilled bicycle mechanic.

Endnotes

1. The Consumer Product Safety Commission (CPSC) has statistical data on this issue. However, due to the fact the accident must be reported to the CPSC, the results are skewed. Emergency room data is more reliable for actual accidents in a certain population group.
2. I have listed some representative texts and authorities in the Bibliography. The European cycling literature, in particular, is an excellent source of this type of information.
3. Adeyafa, B.A. "Determination of the Loads, Deflections and Stresses in Bicycle Frames." A Dissertation, University of Manchester, Institute of Science and Technology, March, 1978. This relatively unknown piece of research has proven to be accurate in all additional laboratory testings by others that I have evaluated.
4. Brian Jobst, The Bicycle Wheel, Avocet Press. I have run the infinite series model that is referenced in this text and find it to be accurate. The failure of a 'bicycle wheel clearly is a function of the loss of tension values. This can occur from lateral forces due to normal riding when the wheel has not been properly stressed and trued.
5. Ibid.

Chapter 3

The Relationship Between Helmets and Bicycle Accident Reconstruction

Introduction

The purpose of this analysis is to define the impact of bicycle helmet use on a bicycle accident reconstruction.

A substantial debate has raged for years within the cycling community on the impact that bicycle helmet use may have on cycling safety. A number of studies have been done in an effort to end the debate. One such study was conducted in Dade County, Florida, utilizing emergency room accident data on cyclists admitted for treatment. The study revealed that, in 86% of 173 fatal cycling accidents, the most serious injuries were to the cyclist's head and neck. This is a landmark study since it goes on to report the autopsy results of these fatalities. The study results are revealing regarding the use of helmets:

    a. All of the fatally injured cyclists received their injuries from collisions with motor vehicles.

    b. For most of the fatally injured (144 cases), the point of contact was the front third of the motor vehicle.

    c. There were 110 skull fractures in the group studied. All skull fractures were associated with brain injury.

The conclusion of the Dade County study was that the use of bicycle helmets would decrease mortality. It was further concluded that cycling helmets should be designed to withstand blunt trauma.

In addition, several studies have investigated the effect that impact acceleration forces have on the human head and neck. The problem with these types of forces is that acceleration and impact are transmitted to the head over a very short period. It is this very short time period during which the forces are transmitted that causes the impact force to increase dramatically. I have tested the transmittal of impact forces to a cyclist's head and achieved a force of impact as high as 30,000 lbs per square inch over a .01 second impact period. I must emphasize that this value is for a peak acceleration and not the average force.

The results of these studies have led to the promulgation of the ANSI, Snell, and ASTM Standards For Cycle Helmets. Prior to the adoption of these standards, many cycling helmets were manufactured without any performance standard. Testing of helmets not certified by ANSI or Snell revealed that the helmets gave very poor protection. Indeed, testing apparatus inside the helmet was broken during testing.

The Cycle Helmet as a Factor in Bicycle Accident Reconstruction

Laboratory test data and accident statistics clearly show that use of a Snell, ANSI, or ASTM approved helmet will dramatically decrease a cyclist's risk of head injury. However, in the reconstruction of cycling accidents, several factors concerning helmets should be considered before determining the cause of the accident.

Conspicuity

As noted in several studies, only certain colors are visible to motorists. Thus, a bicycle helmet, which is a piece of safety equipment, should have a conspicuous color that can be seen by motorists. The color of the helmet is also important because the head is often the only part of the cyclist's body that is visible in traffic. Appropriate helmet colors are fluorescent tones: yellow, highway orange and lime green. Work done by other researchers has confirmed that these colors increase a motorist's perception-distance by as much as 600 feet, or more. Colors such as black, gray, off-white, and camouflage blend into the background and can decrease the reaction time available to the motorist to less than one second.

Ability of the Cycle Helmet to Withstand Impact

There are helmets on the market that have not successfully passed the Snell or ANSI impact test. To determine whether or not a helmet is "ANSI Approved," "Snell Approved," or "ASTM Approved," look for a label inside the helmet that designates its design impact strength:

- For a Snell helmet, the label will read, "Certified by Snell Memorial Foundation, Rhode Island, Sample of This Helmet Has Passed the Snell Standard and is so Certified, Serial No. . . ."

- For an ANSI Helmet, the label will read, "Meets and Exceeds 290.4 ANSI Standard." (The difference between impact tests, in using the Snell versus the ANSI Standard, is negligible for purposes of accident reconstruction.)

• For an ASTM Helmet, the label will read, "Meets and Exceeds ASTM Standard F 1447.93."[1] As of 2001, the helmet industry is still refining this warning sticker and the testing procedure.

The ability of the cloth covered, "soft shell" helmets to withstand impact was seriously questioned in recent accident reconstructions. Analysis of these helmets has shown that they will only withstand one severe blow before shattering. As a result, the second blow in a multiple-blow accident may be life-threatening. Additionally, accident investigations have shown that these helmets do not skid along road surfaces. As a result, the helmet will tend to "hold" to the road surface while the rider's body mass continues along some type of trajectory. Obviously, this can have disastrous consequences for the cyclist's neck. Hard shell helmets will, on the other hand, "skip" along the road surface and, thus, can withstand multiple impacts.

Accident investigations have also shown that helmets can come unstrapped, or the fastening system can fail, during head impact. This type of helmet failure may result from the cyclist's inability to properly fasten/adjust the helmet strap. The failure may also result from an under-designed helmet retention system.

The current Snell and ANSI Standards do address the helmet fastening system and provide design criteria in the form of impact test data. However, the data that I have collected during my accident investigations suggests that more specific testing criteria is needed than that found in the 1989 Snell Standard for helmet fastening. In addition, a specific type of design is needed that will pass an adequate impact test. These issues should be evaluated by the accident reconstructionist on a case-by-case basis. As of 2001, the hemet industry has generated helmet fastening sytems that perform well on impact tests.

Helmet vs. No Helmet in Bicycle Accident Reconstruction

Accident investigations have shown that a Snell, ANSI, or ASTM approved helmet may provide enough protection to take a life threatening situation and turn it into an inconvenience. However, the wearing of a bicycle helmet is not a guarantee of complete protection against injury. A soft shell, cloth covered helmet can actually increase injury if, as discussed above, it adheres to the road surface. A soft shell helmet will also shatter on impact and will not protect against a secondary impact. Moreover, even a hard shell, Snell approved helmet will not protect against impact injuries where certain force angles are a factor. For example, a landing on top of the head (after going over the front of a cycle at a high rate of speed from front wheel disengagement or the equivalent) has resulted in quadriplegia even though the cyclist wore a Snell approved helmet. It is therefore apparent that bicycle helmets and the Snell/ANSI Standards, which standards are only minimum design criteria, cannot protect against all impact injuries.

The Bicycle Accident Reconstructionist and the Helmet Issue

There are several steps the reconstructionist can take to identify the causal factor(s) of a bicycle accident in which helmet use is at issue. A list of these teps is set forth below. Please note that, given the great variety of accident scenarios, the list is, by necessity, not comprehensive. In applying these steps, the reconstructionist should employ sound principles and practices of Engineering:

1. Injuries When a Helmet is Worn

Cyclists often sustain severe spinal injuries even though they wore an ANSI or Snell approved helmet.[2] These injuries have included quadriplegia from the rider going over the front of the cycle and landing directly on top of the helmet. In such an accident, the force of landing is transmitted along the spine, thereby causing the quadriplegia. Of course, a 30,000 lb. force transmitted directly to the head, probably would result in death without the helmet.

I have interviewed many physicians and reviewed many health journals during the course of my accident investigations. This work has shown that there is no real consensus among health care professionals regarding the minimum amount of force that will cause injury when transmitted to the head. However, an Engineering analysis of the issue indicates that the quick deceleration from impact causes a rapid increase in transmitted force to the head. Anything that prevents this transmittal, such as a Snell, ANSI, or ASTM approved helmet, will greatly aid in preventing head injury. Again, it should be emphasized that a rider may sustain catastrophic injuries even though he/she wore a bike helmet. Consequently, the reconstructionist should carefully observe the biokinetics of the accident to determine transmittal of forces.

In addition, if a soft shell helmet is worn, the reconstructionist should investigate the potential for multiple blows to the head. The history of the helmet and the potential for misuse prior to the accident should also be investigated. As noted in the Snell Standards, warnings in the helmet must advise against exposing the helmet to solvents that could damage the lining or the shell. Such chemical exposure can sometimes be identified simply by examining the condition of the helmet.

Soft shell helmets with cloth covers are frequently unable to skid and, as a result, hold the head to the pavement. Again, the biokinetics of the fall should be analyzed, particularly if the accident victim or witnesses is/are available.

2. Injuries When a Helmet is Not Worn

It is frequently alleged that a cyclist was negligent in failing to wear a helmet. However, as indicated above, a cyclist may sustain traumatic injuries that have no bearing on whether he/she wore a helmet. Therefore, in attempting to determine the cause of an accident, the reconstructionist should ascertain, through the medical community and other sources, whether the use or absence of a helmet contributed to the injuries sustained by the cyclist. The absence of a helmet does not constitute contributory negligence unless the Engineering dynamics clearly establish the causal connection to the injuries sustained.

Clearly, the Engineering data and accident statistics show a direct relationship between head trauma and the lack of a helmet. The use of a Snell, ANSI, or ASTM helmet should reduce head injuries by approximately 80%. If a head injury results, the lack of a helmet may be a causal factor. However, in at least one type of fall (where the cyclist goes over the front of the cycle and lands directly on the top of the head) the use of a helmet may actually contribute to spinal injury.

Conclusion

Although it is a safety device, a bicycle helmet may contribute to an accident if it is not conspicuous. A helmet that is black, gray, or another neutral color cannot be seen by motorists and, thus, will reduce driver reaction times. In other words, if a cyclist cannot be seen, the motorist cannot stop in time to avoid the accident. The proper colors for a cycle helmet are fluorescent lime-green, yellow or highway orange.

The use of a helmet does not necessarily prevent injury. The issue of contributory negligence, on which Engineers are often asked to give an opinion, is a function of the biokinetics of the particular type of accident.

Endnotes

1. ASTM. Standard Specification for Protective Headgear Used in Bicycling. American Society for Testing and Materials (Philadelphia, PA). F 1447-93. 1993:1-2.
2. The ASTM standard is, as of 2001, used in the bicycle helmet industry. Testing has shown that the standard is at least the equivalent to ANSI or Snell. However, accident data is not yet available for analysis.

Chapter 4

The Impact of Roadway and Cycle Path Design on Bicycle Accident Reconstruction

Introduction

During the investigation of a bicycle accident, the design of the subject roadway or bike path is often an issue. In these cases, a pure Civil Engineering analysis is required. The Civil Engineering profession has gone to a great deal of trouble to generate design standards for roads and bike paths that will be used by cyclists. Accordingly, this analysis will focus on officially designated bikeways and bike paths. It should be noted that there are no general design standards for roads with foreseeable bicycle usage that are not designated as cycling roadways. This gap in design standards is unfortunate since all roadways should be designed and constructed to be bicycle-friendly. This analysis will also emphasize the application of the well known design standards for bikeways and bike paths to bicycle accident reconstruction.[1,2]

Bicycle Path Design

Bike Paths which are commonly referred to in the Engineering literature as "Class I, Bikeways," serve corridors other than streets, highways, or areas with wide right-of-ways. The paths may therefore be constructed away from the influence of parallel streets. The Engineer must have a clear understanding of cycling behavior and motor vehicle traffic patterns when designing these facilities. In addition, the Engineer must be aware of the path widths, slopes, and curvatures contained in the Engineering literature. These design concepts were painstakingly developed by facility designers with knowledge of the appropriate velocities and handling characteristics of cycles.

Where facility design is at issue, the Engineer should determine whether applicable standards were met during the design and construction of the facility. (Local codes usually rely on the Federal and State Standards for compliance by the designer, developer and construction manager.) The Engineer must then determine if one or more design variances actually caused the accident.

Designated bike paths have a design speed of 20 mph or 30 mph on grades steeper than 4% and longer than 500 feet. This design factor is only one of many contained in the Engineering literature. The specific concern of the reconstructionist is to determine how these design factors were applied. This determination is particularly important with bike paths that are frequently used by young children and other untrained cyclists.

To determine if the bike path was the cause of an accident, the reconstructionist should accomplish the following:
- Survey the accident site to determine the actual site layout at the time of the accident.
- Obtain the "As Built" Engineering Drawings and compare the drawings to the site survey.
- Establish a sequence of events for the actual construction of the facility.

Although the applicable Civil Engineering standards are well known, the failure to implement these standards is a connnon cause of bike way accidents. The following examples illustrate this type of accident reconstruction.

Bicycle Path Construction

A young girl was descending a bike path in a development when she lost control of her bike, was thrown 40 feet, and sustained severe injuries upon impact. Inspection and survey of the path revealed a slope of 38% at the accident site.

The Engineering firm in charge of the design of the facility was named in the lawsuit. Inspection of the Engineering drawings on microfilm at the County offices revealed that the actual slope specified on the "As Built" drawings was 5%. Further evaluation revealed that the contractor, who was supplied with the drawings, completely ignored the designated slope.

Bicycle Path Maintenance

A cyclist veered off of a designated bike path into the path of an oncoming car. Defense counsel contended that the cyclist was at fault since he was off the bike path at the time of the collision.

A traffic flow study of the accident site was conducted. An analysis of 28 cyclists riding the bike path showed that all 28 veered off the path and onto the road in the same location as the injured rider. Further inspection of the path showed that it was impossible to ride on during and after rain storms since water flowed directly onto the path. This water problem was a continuing violation of the Civil Engineering Standards for bike path maintenance, as well as the direct' cause of the accident.

Conclusion

A review of bikeway and bike path cycling accidents reveals that there is a direct relationship between the accidents and the application of the Civil Engineering Standards for designated bikeways and bike paths. To determine if the bikeway or path was a direct cause of the accident, the Engineer must ascertain whether or not the applicable design standards were violated. In addition, the Engineer must carefully consider the other accident factors discussed in this work.

Endnotes

1. American Association of State Highway and Transportation Officials, "Guide for the Development of Bicycle Facilities," prepared by AASHTO Task Force on Geometric Design, Washington, August, 1991.
2. American Society of Civil Engineers, "Bicycle Transportation: A Civil Engineer's Notebook for Bicycle Facilities," 1980, Revised 1994.

Chapter 5

A Determination of Force Onto the Cycle-Rider or Pedestrian During the Impact of a Motor Vehicle

Introduction

During the reconstruction of bicycle accidents, it is often necessary to determine the force of impact onto the rider. This determination can help forensic engineers and medical doctors define the cause of the accident and subsequent injuries. Unfortunately, the Force of Impact is frequently impossible to obtain by measurements. Therefore, the purpose of this mathematical analysis is to provide forensic engineers with a formula for determining Force of Impact.

Development of Force of Impact

Assume Force of Impact is not related to gravity:

Force incurred is due to deceleration, therefore:

$$\text{deceleration} = \frac{dv}{dt} = \frac{\Delta V}{\Delta T}$$

where; ΔV = Change in velocity over deflation of the tire and velocity prior to deflation, minus the final velocity.

$$\Delta T = \text{Time Increment}$$

To determine ΔT,

$$\overline{V} = \frac{dv}{dt} = \frac{\Delta X}{\Delta T}$$

$$\Delta T = \frac{\Delta X}{\Delta V}$$

where;
$\overline{V}$ = average velocity over deflation
ΔX = deflation of cyclist upon impact

ΔT = time increment

To determine Force:

$$F = M \cdot \Delta A$$

where;

M = mass

ΔA = Acceleration converted to deceleration for purposes of reconstructing accidents. Always assume the cyclist is the mass. Thus, if a car traveling at 50 mph strikes a 150 lb. cyclist traveling at 10 mph head-on, assume for analytical purposes that the cyclist strikes the car at 60 mph. The mass of the vehicle is so large in comparison to the cyclist, that the mass of the cyclist is the controlling factor.[1]

To calculate the Force in the above example, the calculations are:

$V_1 = 60 \text{mph} \cdot 1.47 \text{ ft/sec/mph} = 88.2 \text{ ft/sec}$
$V_2 = 0$

where;

V_1 = velocity prior to impact
V_2 = velocity after deceleration

$$\bar{V} = \frac{V_1 - V_2}{2}$$

$$\bar{V} = \frac{(88.2 - 0)}{2} = 44.1 \text{ ft/sec}$$

$$\Delta T = \frac{\Delta X}{\bar{V}} = \frac{0.333}{44.1} = 0.00755 \text{ sec}$$

where;

ΔX = The distance the cyclist's body moves while absorbing the Force.[2]
Assume 4 inches or 0.333 feet.

$$\Delta A = \frac{\Delta V}{\Delta T} = \frac{(88.2 + 0) \text{ft/sec}}{0.00755 \text{sec}} = 11,682 \text{ ft/sec}^2$$

The 150 lb. mass of the rider must be converted to a "slug" to account for the acceleration of gravity on the rider.
Gravitational Mass,

$$G_m = \frac{\text{Mass}(lb_m)}{32.2 \frac{lb_m \cdot ft}{lb_f \cdot sec^2}}$$

$$= \frac{150 \ lb_f \cdot sec^2}{32.2 ft}$$

$$= \frac{4.66 lb_f \cdot sec^2}{ft}$$

Force of Impact = F_i = G_m Deceleration

$$F_i = \frac{4.66 lb_f \cdot sec^2}{ft} \cdot 11{,}682 \ ft/sec^2$$

F_i = 54,438 lbs. force

Conclusion

The above Force of Impact formula assumes that the full force of the impact is absorbed by the surface area that is struck. This area can be one square inch or one square foot as long as the full impact is absorbed by the body at one time. The formula's variables may be changed to reflect actual conditions at the accident site. It should also be emphasized that this formula expresses engineering probability and represents the order of magnitude involved. Different values can be derived by changing variables through laboratory testing and various research sources.

Endnotes

1. I have actually heard "experts" testify, under oath, that the earth moves when a cyclist falls, thus, the mass of the earth must be taken into account to calculate the impact of the collision. This theory is, however, incorrect. Bicycle accident reconstruction involves principles of basic physics, not intergalactic astronomy.
2. Unlike a rigid steel ball bearing, a cyclist is a fairly elastic body. As a result, the cyclist will absorb the force of impact rather than being thrown a great distance. To measure precisely the distance that the cyclist moves, the Engineer should consult the appropriate medical literature.

Chapter 6

Determination of the Reaction Times Available to a Cyclist at Different Intersection Configurations

Introduction

Collisions at intersections represent most car-bicycle accidents. The accidents that I have investigated typically revealed that the motorist did not see the cyclist and/or the cyclist did not see or expect the car to turn into the cyclist's path. This study, based on simulations and field investigations, seeks to analyze this common accident scenario by examining the turning of a single passenger car at various intersection configurations. The information revealed by the examination may, as here, be used to determine the reaction time available to the cyclist by plotting the trajectory of the cyclist's turn until impact. Since the motorist's ability to see the cyclist (and react) is a conspicuity issue, any conclusions based on conspicuity will be discussed with other conspicuity issues later in the book. Thus, the focus of this simulation study is on the ability to predict the time required for the car-bicycle collision, given the physics of the event at various intersection designs. Again, it should be noted that, where possible, simulation values were verified by actual field measurements.

Analysis of Different Intersection Configurations

As illustrated in Figures 1 through 5, different intersection configurations were used in the simulation. A Category 4 passenger car (e.g., Chevrolet Caprice or Ford Thunderbird), as defined by the EDSVS simulation modeling program,[1] was assumed. The variables used in the model include vehicle class category, tire/ground friction coefficient, vehicle weight and velocity. The different intersection configurations studied were:

- Figure 1 - Two lane, four-way intersection - In this simulation, a car turns left from a right lane to another right lane in a total two-lane configuration. All four input sections of the intersection contain two

lanes. The car crosses the path of the cyclist as the cyclist attempts to turn right.
- Figure 2 - Four lane intersection - In this simulation, the input sections of the intersection are each four lanes. The car turns left from the right lane into the right lane of the receiving lane. The car passes in front of the cyclist. The cyclist is turning from the right lane into the same right lane as the car.
- Figure 3 - Four lane turn into two lanes - This simulation has a car turning from the right lane of a four-lane roadway into the right lane of a two-lane roadway. The cyclist is turning from the right lane of the four lane road into the right lane of the two-lane road. The car passed in front of the cyclist.
- Figure 4 - Vehicle turns right, cyclist turns left from four-way intersection onto a two-lane road - This simulation has the car turning from a left turn lane into the right turn lane of a two-lane road. The cyclist approaches the intersection in the left lane and turns left into the right lane of the two-lane road. The simulation illustrates that the car travels throughout the radius of the turn in 3.24 seconds. This is the same amount of time that the cyclist has to avoid the accident.
- Figure 5 - Vehicle turns right from a two-lane road into the left lane of a four-lane, one way road, cyclist turns left into the far left lane of four-lane road - This simulation assumes that the cyclist violates the rules of the road and enters the one way intersection in the far left lane. The car travels through the radius of the turn in 3.36 seconds. Therefore, the cyclist has 3.36 seconds to avoid the accident.

To determine the reaction time available to the cyclist, the bicyclist was positioned prior to initiating the turn) in accordance with the rules of the road. The positioning of the cyclist also took into account the riding habits of an experienced cyclist and an assumed working 10-speed road bike.

The trajectory of the cycle to a point of impact on the car was plotted. Then, using cycle velocities of 10 mph and 15 mph, the cyclist's reaction time was calculated until the point of impact on the car. These reaction times are noted in Table 1 and Table 2. The reaction times available to the cyclist varied from 1.62 seconds to 3.94 seconds for 10 mph and 1.08 seconds to 2.62 sec for 15 mph (Table 1 and Table 2).

Table 1 - A Definition of the Reaction Times and Time of Travel of a Class 4 Vehicle and a Cyclist at Different Intersection Configurations

Intersection	Time of Travel of Car	Reaction Time Available to the Cyclist (Table 2)	
		@ 10 mph (secs)	15 mph (secs)
Figure 1, 2-lane, 4-way intersection	3.38 secs	1.62	1.08
Figure 2, 4-lane	3.30 secs	2.18	1.45
Figure 3, 4-lane turn into 2 lanes vehicle, left lane into right lane, cyclist, right lane into right lane	2.93 secs	1.91	1.27
Figure 4, 4-lane into 2-lane intersection; vehicle left lane into right lane, cyclist left lane into right lane	3.24 secs	3.94	2.62
Figure 5, 2-lanes into 4 lanes; vehicle right lane into left lane, cyclist, right lane into far left lane	3.36 secs	3.54	2.35

Table 2 - Distance, Velocity, and Reaction Time Available To Cyclist for Five Different Intersection Configurations

Figure	Distance Cyclist Travels Prior to Impact	Cyclist Speed 10 mph (ft/sec)	Cyclist Reaction Time @ 10 mph (sec)	Cyclist Speed 15 mph (ft/sec)	Cyclist Reaction Time @ 15 mph (sec)
1	23.8 ft	14.7	1.62	22.1	1.08
2	32.0 ft	14.7	2.18	22.1	1.45
3	28.0 ft	14.7	1.91	22.1	1.27
4	58.0 ft	14.7	3.94	22.1	2.62
5	52.0 ft	14.7	3.54	22.1	2.35

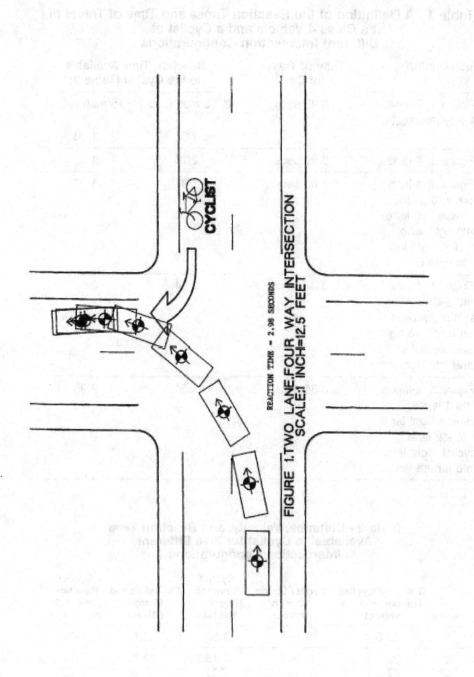

FIGURE 1. TWO LANE, FOUR WAY INTERSECTION
SCALE: 1 INCH = 12.5 FEET
REACTION TIME = 2.96 SECONDS

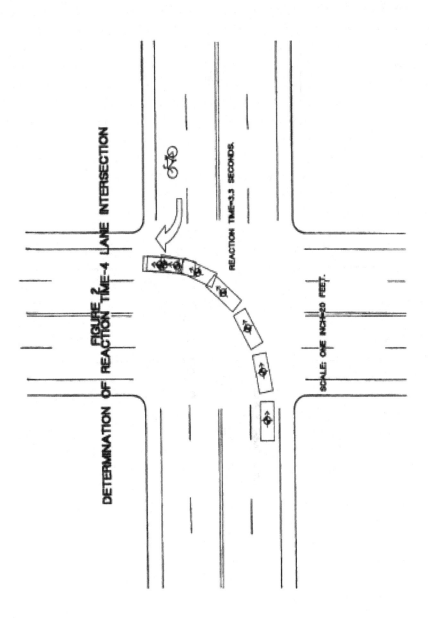

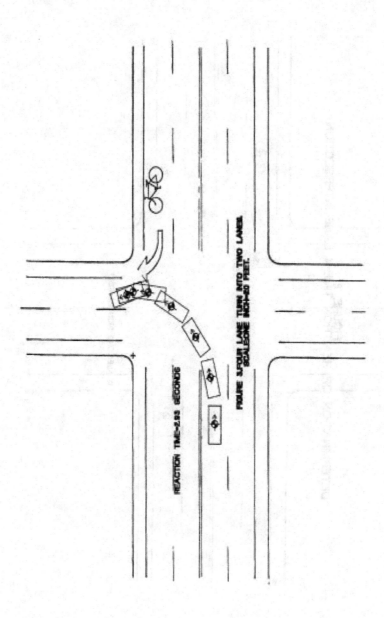

Determination of Reaction Times at Intersections

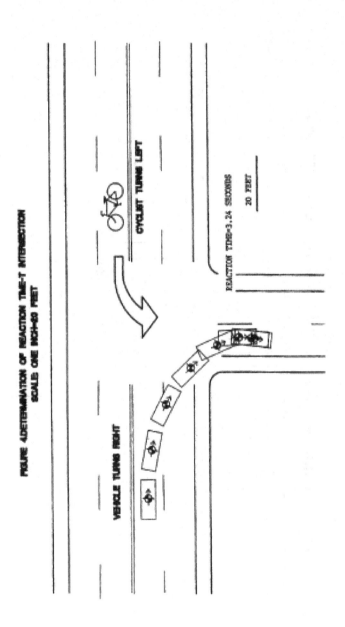

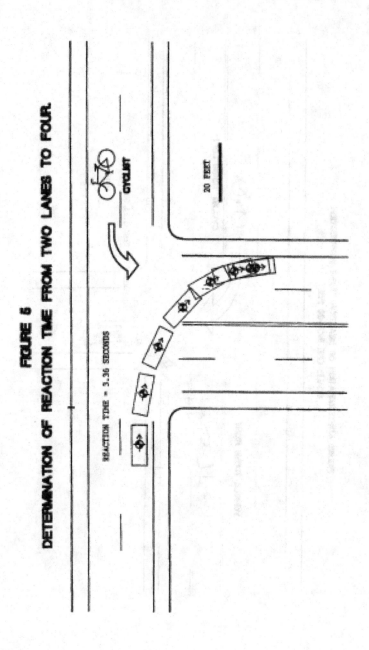

FIGURE 6
DETERMINATION OF REACTION TIME FROM TWO LANES TO FOUR.

Conclusion

Using the reaction times available to the motorist and cyclist at the five basic intersection configurations, as well as a basic surprise reaction time of 1.0 seconds, the following conclusions can be made:

1. When traveling through an intersection, a motorist has a greater reaction time than a cyclist. The car's time of travel ranges from 2.93 to 3.36 seconds, while the cycle's time of travel prior to impact ranges from 1.08 to 3.94 seconds. The only time the driver's reaction time is less than that of the cyclist is when the motorist is turning illegally or irresponsibly (see Figures 4 and 5).
2. Given the reaction times available to the motorist and cyclist, the burden of impact avoidance is on the motorist. If the motorist and cyclist proceed through the intersection simultaneously, the cyclist is very close to the limits of the reaction time available for safely avoiding the accident. The motor vehicle will have a slower radius of turn due to the engineering forces involved than will the cyclist. As a result, the motorist has a greater time to avoid the accident than does the cyclist.
3. The overall conclusion of this study is that, once a cyclist is committed to a turn at an intersection, he/she no longer has the reaction time required to avoid an impact. However, the motorist still has time to react and avoid the collision. In light of this conclusion, a cyclist should always be aware that, once he/she enters an intersection, every motor vehicle is potentially capable of hitting him/her.

Endnotes

1. Author's Note: Various vehicles were used in the field to verify the turning time as simulated by the EDSVS Category 4 vehicle. These different vehicles were found to exhibit the exact turning speed as predicted by the EDSVS model. As a result, the EDSVS model was used to calculate driver reaction times. Additional information on the EDSVS accident reconstruction model is available from: Engineering Dynamics Corporation, 8625 SW Cascade Blvd., Suite 200, Beaverton, Oregon 97008-7100.

Chapter 7

A Determination of the Causal Factors of Bicycle Accidents Caused by Front Wheel Release and Measures to Prevent These Accidents

Introduction

My investigative work has shown that one of the most prevalent bicycle accidents involves the premature release of the front wheel.[1] When a bicycle's front wheel prematurely releases, the rider is suddenly thrown forward and, as a result, may sustain catastrophic injuries. The purpose of this analysis is to define the causes of this type of accident and the available preventative measures. Included in this analysis are the biokinetics of a cyclist's fall, as defined by data from actual accident investigations.

Analysis of Falls Caused by Premature Release of the Quick Release Mechanism

A quick release mechanism permits the fast and easy removal of a bicycle wheel. The mechanism consists of an axle with a nut on the right side, as viewed by the rider, and a flange nut on the left side. To secure the wheel during riding, the flange nut is forced into a locked position which causes the ridges on the nut to emboss, or force regular abrasions on, the sides of the front fork (See Figure 1).

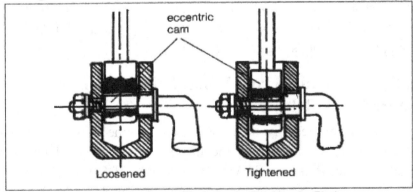

Figure 1 Principle of quick-release. (Illustration by Rob Van der Plas)

Quick release mechanisms are located on both the front and rear wheels. However, a premature release of the rear wheel does not pose the risk of catastrophic failure and injuries that premature front wheel release does. Consequently, this analysis examines only the premature release of the front wheel.

When the front wheel clamp is tightened, the torque value on the axle nut must be greater than 5 foot-pounds and less than 30 foot-pounds. If the torque value exceeds 30 foot-pounds, the axle may break as a result of the unusually high force value. Experienced bike racers often over tighten quick release mechanisms prior to racing on difficult terrain such as cobblestones or off-road. However, I have only encountered axle breaks from over tightened quick release mechanisms twice during many years of professional practice.

It is far more common for a quick release failure to occur when the front wheel axle nuts are not tightened to 5 foot-pounds. When this occurs, the wheel can prematurely release simply by the rider shifting weight toward the rear during normal riding or prior to encountering some road change. It must be emphasized that 5 foot-pounds of torque is a generalization. The minimum acceptable torque value may be as low as 1 foot-pound, depending on the axle and front fork design. The important point about the lower value is that embossing of the front forks must occur in order for the wheel to remain in the front dropouts during normal riding.

The lower design values for quick release mechanisms may differ, and thus impact the risk of premature wheel release, when the front fork dropouts are painted. In laboratory tests, lower design values for quick release mechanisms on unpainted front forks ranged from 0.8 to 5.0 foot-pounds. The lower design values for the mechanisms on painted front forks ranged from 1.7 to 5.9 foot-pounds. These laboratory results show that, with painted front forks, front wheel release occurs at a higher torque value. In other words, the quick release mechanism must be torqued higher to retain the wheel when the front forks are painted. This is due to the fact that the paint impedes the quick release mechanism from properly embossing the front forks.

Preventing Premature Front Wheel Failure From Premature Quick Release

Many devices have been invented for preventing the premature release of quick release mechanisms. Of all the devices tested in my laboratory, the Quick Release Hub Retention Device by Brilando of Schwinn is by far the most effective (See Figure 2, A-D). Even with the application of over 5,000 pounds of force and a torque value on the fastening hub of less than 0.1 foot-pounds, the wheel with the Brilando device did not release.

If only one positive retention device, such as the Brilando, is in place, the wheel will release at .approximately 500 pounds of force from vibration or side to side movement. However, when positive retention devices are in place on both sides of the axle, the front wheel will not prematurely release under normal riding conditions. It must be cautioned that bicycle shops may remove a positive retention device in order to install a velometer on the front wheel.

It has been my experience that, if either of the positive retention devices was in place, premature wheel release was not the cause of the accident. However, if the devices are not in place, premature front wheel release cannot be eliminated as a causal factor.

The best way to prevent premature front wheel release is to not use a quick release mechanism. The quick release mechanism was designed to aid bicycle racers in the quick removal of the wheel under racing conditions. Accordingly, the mechanism should only be used by experienced cyclists who have received adequate training in its proper use. All other cyclists should ride bikes with a closed end front fork assembly ("Closed Assembly"). The Closed Assembly, which has been available since 1896 (see Figure 3), prevents the potentially disastrous failure that can occur with the quick release design.

Biokinetics of a Fall From Premature Release of a Quick Release Mechanism

In reconstructing a rider's fall due to the premature release of a front wheel, knowledge of the biodynamics of the fall is important for determining whether front quick release failure actually caused the accident. Several studies have been conducted in the European and American professional cycling communities on the physiology of cycling.[2] In addition, the engineering forces involved in cycling have been studied exhaustively by Jim M. Papadopoulos.[3] These studies reveal the following general biokinetic facts relevant to quick release accidents:

- Assuming level ground, if the front wheel completely releases due to failure of the quick release mechanism, the front forks will strike the ground first. As a result, the dropouts on the front forks will be traumatized from contact with the road. The front forks are typically bent either into the frame or severely forward, depending on the rake angle of the front forks prior to the accident. In this type of accident, the rider does not lose balance prior to the wheel releasing. Rather,

the front wheel releases when the rider shifts his/her weight toward the rear or shifts his/her center of mass during normal riding motion. When this occurs, the front forks drop onto the asphalt and the rider is thrown over the front of the bike.
- Assuming level ground, if the front wheel "hangs up" on the front forks, the forks drop onto the wheel axle and momentarily halt their descent. Or, the bottom part of the crown can land directly on top of the wheel if the forks do not "hang up" on the axles. Both types of failure may cause the rider to lose balance and fall either to the side or over the handlebars. In these types of falls, the rider usually strikes the ground before the front forks. This is due to the fact that the rider's upper body is, as a result of lost balance, lower than the front fork dropouts. It is thus apparent that premature front wheel release does not always yield front forks bent from trauma.

General Reconstruction of Accidents Due to Premature Release of Quick Release Mechanisms

If an Attorney or Forensic Engineer encounters a cycle accident involving an issue of premature front wheel release, he/she should be aware of and/or observe the following investigation process:

- Interview witnesses to the accident. If the front wheel was observed releasing prior to the fall, then the cause of the accident was likely premature wheel release.
- Carefully inspect the cycle to see if the original equipment is still on the cycle. It is possible that something as critical as the front forks was changed, thus altering the configuration of the system. Different equipment can change the design limits of the quick release mechanism.
- Determine whether or not a positive retention device was removed by the cyclist or a bike shop. Some cycle companies include positive retention devices as standard pieces of equipment. However, I have seen at least one instance where a positive retention device was removed by a bicycle shop in order to install a velometer. The removal of the device, which was provided by the cycle adequate warnings, impeached the integrity of the manufacturer with system.

As indicated above, use of positive retention devices or closed loop fork ends can prevent premature release of quick release mechanisms. Accordingly, investigation of premature front wheel release accidents should focus on whether a positive retention device was in place. If witnesses were present, emphasis should be given to the specific rider dynamics of the accident. The causes of the accident can then be determined from the markings and damage to the cycle.

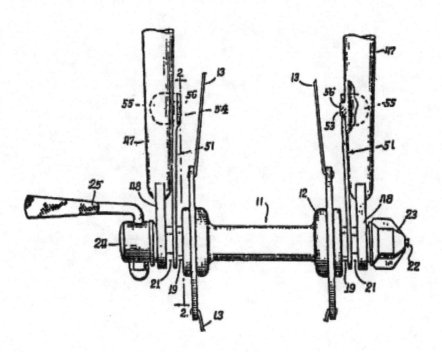

Figure 2 Front Fork Positive Retention Device – Brilando Patent

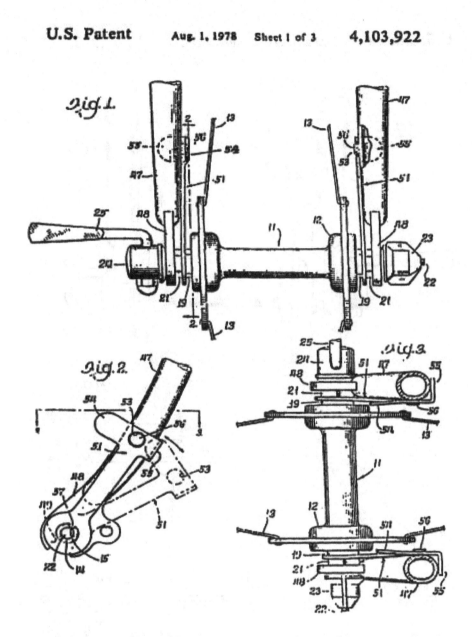

Figure 2 Front Fork Positive Retention Device – Brilando Patent

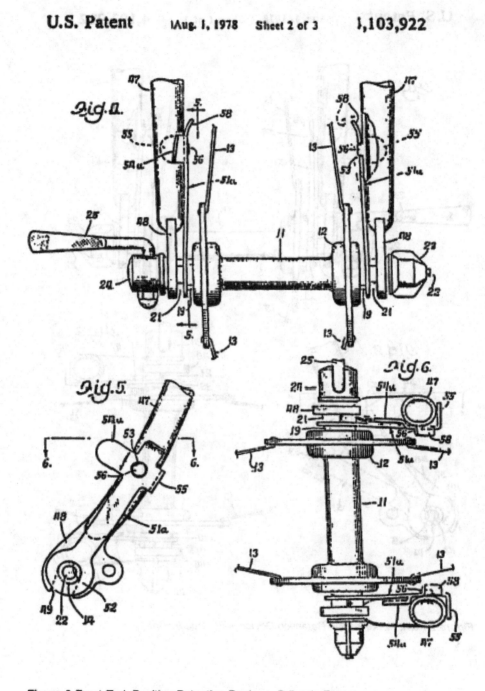

Figure 2 Front Fork Positive Retention Device – Brilando Patent

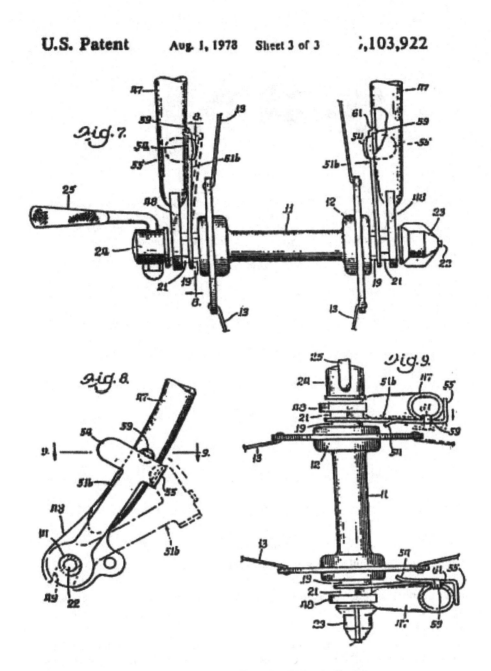

Figure 2 Front Fork Positive Retention Device – Brilando Patent

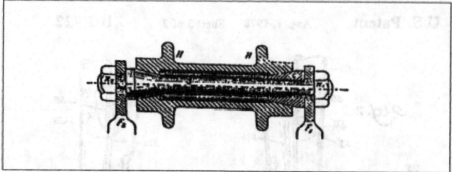

Figure 3 The design of a Closed End Front Fork Assembly for bicycles. Figure 392, page 369, *Bicycles and Tricycles, An Elementary Treatise on Their Design and Construction*, by Archibald Sharp. Published in 1896, available through MIT Press, Massachusetts Institute of Technology, Cambridge, Massachusetts, 02142.

Endnotes

1. Various bicycle manufacturers are now installing positive or redundant retention devices on their bikes. The increased use of these devices has, happily, led to a dramatic decrease in premature front wheel release accidents. The industry, thus, appears to be well on its way to completely eliminating the type of accidents that are the subject of this analysis.
2. Faria, Irwin E. "Cycling Physiology for the Serious Cyclist," California State University, Sacramento, California, Charles Thomas Publisher, 1978 - Several other publications are in the literature on this subject.
3. Papadopoulos, Jim M., "Bicycle Steering Dynamics and Self Stability: A Summary Report on Work in Progress," Cornell Bicycle Research Project, Preliminary Draft, 12/15/87. This is the definitive work in this area. My major critique of this effort is the lack of field verification. Most of the conclusions I have verified in the field, although not quantitatively.

Chapter 8

A Determination of the Casual Factors in Bicycle Accidents at Railroad Crossings

Introduction

The purpose of this analysis is to define the causes of bicycle accidents at railroad crossings. Bicycle accidents at railroad crossings usually result in a fall with serious upper body injuries to the cyclist. This paper excludes cyclist collisions with a train or motor vehicle. Rather, emphasis will be placed on the following cases of railroad crossing bicycle accidents:

- Front wheel disengaging while crossing the railroad tracks.
- Improper design or maintenance of the tracks.
- Weather conditions or cyclist error.

The impetus of these three types of accidents is examined below. Please note that the calculations of the force with which a cyclist's body moves against resistance (impetus) are based on actual field investigations or accident reconstructions.

Front Wheel Disengaging

When a front bicycle wheel is not properly tightened, the wheel is likely to disengage during a railway crossing as a result of a shift in the cyclist's weight to the rear of the bike.

Typically, a standard road bike has a weight distribution of 60% on the rear wheel and 40% on the front wheel. Note that this weight distribution will vary for triathlon cycles with time trial bars. With such cycles, the weight distribution may be skewed to 60% on the front wheel and 40% on the rear wheel. A cyclist will, however, redistribute his/her weight to the rear in anticipation of a railway crossing. This weight shift places the rider/bicycle system at risk if the front wheel is not properly tightened. When the front wheel has disengaged during a crossing, an inspection of the cycle will indicate which of the following falls occurred:

- The front forks strike the ground before the cyclist - In this instance, the trauma to the front forks is obvious. The forks are bent either forward or toward the rear of the cycle. The road bed around the railroad tracks will also show trauma.

- The cyclist strikes the ground before the front forks - When the front wheel disengages after the rider shifts weight, the crown may land on the tire, thereby causing the front forks to lock behind the front axle. Since this acts as a brake to the rider/cycle system, the cyclist can fall to the side or over the handlebars, depending on the cycle's speed. In this type of fall, the cycle will exhibit very little damage to the fork area. However, the front forks may show some bending toward the rear, depending on the strength of the forks.

In the 1970's, Schwinn recognized the serious risk of premature front wheel release on bikes equipped with a quick release mechanism. The solution to this problem was the Brilando patent for a positive retention device. (The Engineering dynamics of premature front wheel release are discussed elsewhere in this book.) Thus, where it is alleged that premature front wheel release caused the subject railroad crossing accident, the investigator should determine whether or not a positive retention device was in place on the front fork. If this question is answered affirmatively, the investigator may rule out premature front wheel release as a cause of the accident.

Premature front wheel release may also be ruled out as an accident causal factor if the cycle has a closed end front fork. This fork design, which has been available since the 1890's, does not allow wheel release unless the axle is removed from the front dropouts.

Improper Design or Maintenance of the Tracks

A bicycle is a one-track vehicle with certain balance characteristics built into the rider/cycle system. This system can be unbalanced by any unusual protuberances or abrupt changes in the road surface at railroad crossings. Civil Engineering standards have been promulgated for the design and maintenance of railroad crossings which seek to insure cyclist safety. Compliance with these standards, which are used in the American Railway Engineering Association (AREA) Guidelines for Construction or Reconstruction of Highway Railway Crossings (see Appendix), will prevent unexpected road surface changes that can cause a cyclist to fall. Examples of Engineering Drawings that reflect the AREA standards are noted in the Appendix (Figures 1 through 4).

Violation of the above Engineering standards may result in several conditions that can cause a cyclist to fall at a railroad crossing. These conditions are as follows:

- Free Rail - Either during construction or through normal wear and tear, one or several rails can loosen from the standard cross ties. When this occurs, the rail will "bounce" or "float" as much as several inches above the level of the street when motor vehicles pass over the crossing. A cyclist crossing the tracks at the same time as a motorist can be thrown from his/her bike by this type of "free" rail. This phenomena generally results from the railroad's failure to maintain the crossing in accordance with AREA standards.
- Uneven Surface - Crossing timbers and drive spikes should be even with the road surface and rail. Uneven surfaces between the crossing timbers and rail can cause an unexpected shift in rider weight that results in a fall. Additionally, air spaces on both sides of the rail should be no greater than 2.5 inches. Larger air spaces will not permit safe passage of a standard 27 inch bicycle wheel. In such a situation, the air space-rail arrangement literally grabs the front wheel of a cycle that is ridden in a perpendicular path over the rails.
- Transition Between Road Surface and Crossing Timbers - The responsibility for smooth transition between the road surface and crossing timbers is often ignored. This may be due to a lack of clearly defined areas of design and maintenance responsibilities between the railroad and municipal authority. Nevertheless, if the transition between the two infrastructures is not smooth, a cyclist may lose his/her balance due to unexpected weight shifting while crossing this interface.

Many combinations of the above violations of the AREA standards are possible. Any of these combinations can place a cyclist at risk and cause an accident. If, however, the AREA standards are met at a crossing, the accident may be attributed to rider error or weather conditions.

Weather Conditions or Cyclist Error Resulting in a Fall

Many cycling accidents that occur at railroad crossings are due to rider error or weather conditions. Inexperienced and untrained cyclists are, of course, at greater risk for such accidents. An examination of the relevant engineering drawings (see Appendix) reveals several dangers that exist at railroad crossings for unwary cyclists. These dangers, which may be present even if the crossing complies with all applicable AREA standards, are as follows:

- Lack of Perpendicular Line of Travel - Due to the variety of railroad wheel gauge sizes in the United States, airspace of 2.5 inches is needed between the crossing timber and rail. If the cyclist's line of travel is at an angle such that the width of the bike wheel can drop into this opening, the crossing timber/rail structure literally grabs the bike from under the rider. If this occurs, the rider will fall forward and possibly sustain severe injuries.

- Wet Rails - Rain or moisture from high humidity and condensation can make rails extremely slick. Tests run on wet rails yield coefficients of friction (COF) of ice. Consequently, rails that are easy to ride over during the daytime can become extremely slippery as the temperature drops at night.
- Failure to Redistribute Weight - Although a railroad crossing may adhere to the AREA standards, a cyclist must alter his/her normal riding technique in order to safely negotiate the crossing. When approaching a railroad crossing, a cyclist should slow down, enter the crossing perpendicular to the angle of tracks, and distribute as much body weight over the rear wheel as possible. This technique will give the cyclist as much stability as possible while riding over the crossing. Unfortunately, this recommended riding technique conflicts with the technique now employed by triathletes and other cycle racers. The racing technique distributes an inordinate amount of weight over the front wheel in order to achieve aerodynamic efficiency. Values as high as 80% weight over the front wheel have been found on some triathlete time trial bicycles. This riding technique violates the classic road bike weight distribution of 60% over the rear wheel and 40% over the front wheel. Even when entering a railroad crossing properly, a cyclist is at extreme risk with the weight distribution of the racing technique.

Chapter 9

The Utilization of the Principles of Conspicuity in Pedestrian and Bicycle Accident Reconstruction

An Introductory Discussion

There are many fine physics textbooks available that describe, on an academic level, the principles of conspicuity (the ability to be seen or observed under ambient lighting/color conditions). These academic principles must be applied when addressing issues of conspicuity in an accident reconstruction. Unfortunately, many reconstructionists fail to accurately make the connection between the physics of light and color and the application of these principles to actual reconstructions in the field. I have been present at court and deposition testimony where "experts" have given opinions under oath that would make a high school physics student grimace in disbelief. Where possible, I have included actual testimony on these issues in order to illustrate the misuse of this branch of science.

In light of the above-mentioned misuse/misunderstanding, it is the purpose of this chapter to accurately explain the principles of conspicuity as they apply to the reconstruction of pedestrian and bicycle accidents. Although conspicuity is particularly an issue at night, the ability to perceive, react and avoid a hazard is also important during the day. The concepts developed in this book can, therefore, be applied over an entire 24 hour time period. After the principles of conspicuity are defined, examples of how to properly evaluate an accident site are given.

Properties of Light As Related to Bicycle and Pedestrian Accident Reconstruction

In order to adequately understand the role of conspicuity in pedestrian and bicycle accident reconstruction, certain properties of light must first be explained. The purpose of this analysis is to only define those properties of light which are of interest to the forensic engineer. For this reason, the pure physics of light, which can become very complex, is not discussed here. Rather, those properties of light which can be directly applied to quantifying the causal factors of accidents are the subject of this discussion.

Electromagnetic wave velocity

As Figure 1 illustrates, the description of light as an electromagnetic wave is quite apt. Light is, in essence, an electromagnetic field with electric and magnetic waves travelling perpendicular to one another. More importantly, the waves travel at a speed of 300,000 kilometers per second. At this critical speed, mutual induction continues indefinitely, with neither a loss or gain in energy.

If an electric charge is set into motion within a frequency range of 4.3×10^{14} to 7×10^{14} vibrations per second, the resulting electromagnetic wave will activate the retina of the eye. Light is then generated by the filament of a lamp within this frequency range and seen by the "electrical antennae" of the eye.

As previously indicated, electromagnetic waves, including light waves, travel at a speed of 300,000 kilometers per second (or 186,411 miles per second or 11,184,680 miles per hour). Since the speed of electromagnetic waves is so great, as well as identical, the speed of light is not a factor in calculating reaction times for an accident reconstruction. For example, in the typical surprise reaction scenario, approximately three seconds will transpire from the time a surprise stimuli is perceived by the retina of the eye and the time the individual reacts to avoid the danger or stimuli. In that period of time, a light wave will travel approximately 559,233 miles!

Although their speed is the same, electromagnetic waves have historically been separated into a spectrum of different classes. This classification is called the electromagnetic spectrum.

Electromagnetic spectrum

The electromagnetic spectrum covers the electromagnetic waves that extend from radio waves to gamma rays. All electromagnetic waves, including light waves, differ principally in frequency and wave length.

Different frequencies result in different wave lengths. As anyone who has ever made waves on a water surface has seen, an increase in frequency means a decrease in wave length. Conversely, a low frequency produces long wavelengths. Frequency has historically been measured in hertz.

The relationship between frequency and wave length is defined as follows:

$$c = f\lambda$$

where:
  c = constant speed of light = 300,000 km/set
  f = frequency
  λ = wave length of light

If frequency is defined as a hertz, then one vibration per second, or one electric charge oscillating once per second, is equal to one hertz. The wave length can be calculated as follows:

$$300,00 \text{ km/sec} = (1) \cdot (\lambda)$$
$$\therefore \lambda = 300,000 \text{ km}$$

The electromagnetic spectrum spreads over a very large range of frequencies. The visible region that is of interest in the reconstruction of pedestrian and bicycle accidents lies in the range of:

$$\lambda = 3.8 \times 10^{-7} \text{ to } 7.8 \times 10^{-7} \text{ M}$$
$$f = 3.8 \times 10^{14} \text{ to } 7.9 \times 10^{14} \text{ Hz}$$

Factors Influencing The Visibility of Light

In a vacuum, light is visible to the eye for virtually an infinite distance. In fact, astronomers talk in terms of light years of distance. This is the number of years that light must travel to be seen. An example of this phenomenon is the night sky with its constellations of stars.

Light is not, however, generally observed in a vacuum. Consequently, it is not visible for an infinite distance. Instead, the atmosphere and the curvature of the earth cause the range of light to be visible only up to certain distances. Given the limited visibility of light, it is understandable that the Uniform Motor Vehicle Code, a code commonly referred to by forensic engineers during accident reconstructions, requires motor vehicle headlights to illuminate the roadway ahead for 600 feet. It is also understandable that state codes almost uniformly require that a cyclist riding at night have a bike head light that is visible for a similar distance. Unfortunately, experience has shown that many experts have trouble comprehending these lighting requirements.

In the accident which was the subject of the New Jersey case, a cyclist, Johnson, descended a hill toward an intersection at 12:30 a.m. He rode on the street's double center line at speeds in excess of 40 mph. The cyclist was clearly illuminated by the beams of light from an approaching Jeep's headlights. The Jeep turned in front of the cyclist, thereby causing the cyclist to strike the right rear of the motor vehicle. The cyclist sued the bicycle company on the theory that the company should have attached a bike light to the cycle (in essence forcing the cyclist to buy the light). The plaintiff would not likely have prevailed with this unreasonable theory of liability if his expert had acknowledged the visibility provided by the Jeep's headlights. The expert instead chose to ignore the fact that the Jeep's headlights were required by state statute to illuminate the roadway ahead for 600 feet. The cyclist was, therefore, visible for at least 600 feet prior to the point of impact. Unfortunately, the court did not require plaintiff's expert to answer questions about these statutory visibility requirements.

As the foregoing example reveals, a forensic engineer who is reconstructing a nighttime cycle or pedestrian accident must take into account the zone of illumination that extends at least 600 feet in front of a motor vehicle. The visibility of an object, such as a cyclist or pedestrian, within this zone is a function of the object's reflectivity and the type of colors on the object (or worn by the individual).

Blomberg studies

In an attempt to quantify the recognition distances of pedestrians and cyclists at night under the white light emitted from a motor vehicle's headlights, a series of field tests were conducted by Richard D. Blomberg.[2] The purpose of Blomberg's excellent work involving the effect of white and colored light can be summarized from the abstract of the work:

> A field experiment was conducted to determine the extent of conspicuity enhancement provided pedestrians and bicyclists of at night by various commercially available retroreflective materials and lights. The conspicuous materials were designed to be worn or carried by the pedestrians and bicyclists. Detection and recognition distances for the various experimental and baseline conditions were determined using subjects driving instrumented vehicles over a predetermined route on a realistic closed-course roadway system. Field experimenters were used to model the conspicuity-enhancing materials employing natural motion associated with walking and bicycling. Comparisons of the detection and recognition distances suggested that pedestrians and bicyclists can greatly enhance their conspicuity to drivers at night by wearing certain types of apparel and by using devices that are currently available in the marketplace. Nevertheless, it was concluded that nighttime pedestrian and bicyclist activity is inherently dangerous, even with these devices, and should be avoided.

Blomberg's studies show that the determination of available reaction and perception time at night for an accident reconstruction is a function of the type of clothing and other conspicuity-enhancing materials.

Color and frequency

Color, which is a function of frequency, is often an extremely important factor in forensic engineering evaluations of pedestrian and cycling accidents. Except for light sources, most objects reflect light (from their color) rather than emit light.

The atoms and molecules in objects, such as reflective clothing or a retroreflector lens, have their own natural frequencies. For example, a blue lens on a police car only emits blue light since the lens only has the blue frequency. The other colors in the spectrum are absorbed by the lens. This absorption removes the other colors and, depending on the material, can become warm from this phenomena.

Conspicuity and color

For many years, safety professionals arbitrarily assumed that red was the most conspicuous color in the visible spectrum. Professional cyclists, on the other hand, have known since the mid-1930's that yellow-green or fluorescent yellow-green is the most visible color. If one plots a graph of brightness versus frequency, it can be seen that sunlight is brightest in the yellow-green region. Also, if the development of clothing for professional cyclists is carefully examined, as I have done, it is apparent that there has been a gradual trend toward fluorescent yellow-green coloration. Serious cyclists know that a contest between themselves and a 4,000 pound automobile will usually result in only one winner. Being seen is, therefore, very important to them. Cyclists' desire for self-preservation has resulted in this color trend in cycling clothing.

Many studies using various types of clothing on human subjects have classified certain colors as perception enhancers[4]. It is not surprising that these colors are fluorescent yellow, yellow-green and lime green. A knowledge of a cyclist's or pedestrian's clothing and any perception enhancement devices employed is, therefore, absolutely essential for an accurate determination of available perception or reaction times in an accident reconstruction.

An Evaluation of Bicycle Retroreflector Systems and Their Effect on the Reconstruction of Bicycle Accidents

The bicycle industry meets federal conspicuity requirements by complying with the Consumer Product Safety Commission ("CPSC") Standards for the manufacture of bicycles. These standards require that retroreflectors be placed on the head set, rear of the saddle, and both wheels (two each). Again, all bicycle manufacturers currently comply with these standards.

The effectiveness of the CPSC mandated retroreflector system was questioned at the previously mentioned Johnson trial. In the Johnson accident, the cyclist was dressed in dark clothing, did not have a light on the bike, and was not wearing a helmet. The cyclist was, however, illuminated by the beams of light from an approaching Jeep's headlights. The plaintiff contended that the presence of a light on the bicycle would have prevented the accident. The plaintiff further contended that (contrary to statute) it was the responsibility of the cycle manufacturer, not the cyclist, to attach a bike light to the cycle. Since the bicycle was equipped with a CPSC approved retroreflective system and the cyclist was in the beams of the Jeep's headlights for 600 feet prior to impact, the effectiveness of the retroreflectors was also at issue.

The conspicuity data developed in the reconstruction of the Johnson accident should be quite useful to forensic engineers reconstructing accidents involving conspicuity issues and a cycle with retroreflectors.

The purpose of the next section is to define the capabilities of the CPSC approved retroreflector system in terms of a motorist's perception at night.

Engineering analysis of the retroreflector system

A retroreflector lens consists of hundreds of small mirrors arranged at approximately 90 degree angles to each other. Therefore, when a beam of light hits one of the small mirrors, it is reflected to an adjoining mirror and sent back toward the source. Since these small mirrors are located throughout the lens, a beam of light is not only reflected back at the source but is also dispersed at angles dictated by the angles of the small mirrors. The CPSC requirements for retroreflectors dictate that the total angle of the mirrors shall be 100 degrees. Due to this requirement, light beams from an automobile's headlights that strike a bike retroreflector will be reflected back towards the automobile in a cone with an outer limit of 100 degrees. A typical retroreflector is detailed in Table I. This is the retroreflector that was on the cycle in the Johnson case. To analyze a retroreflector, the CPSC requires that the measurement of retroreflectivity be conducted in accordance with ASTM Standard ES09-91.

The results of a typical retroreflector analysis utilizing the ASTM standards are shown in Table I. Note that the color analyzed is white. The other two colors used on retroreflectors are red and amber. To understand the analysis, the Table should be examined as follows:

- The test distance is 100 feet and the unit of measurement employed by the photometer is candlepower per feet times a centimeter, or cp/ft-c.
- The observation angle is the distance from the norm of the light beam being recorded. This angle is a matter of the test apparatus geometries and is, therefore, always the same. Consequently, the angle of observation is not an issue in our analysis.
- The entrance angle is the same as the exit angle. For example, a source lamp at an angle of 50 degrees to the left of the center axis is required to yield a reading of 6.0 cp/ft-c at a reading site on the photoreceptor. Examination of the measured value yields 9.41 cp/ft-c, which is well in excess of the mandated 6.0 cp/ft-c.
- The applicable standards specify that the most effective angle of entrance and exit for the light source is 10 degrees from the center axis.
- An examination of Table I shows that the analyzed Cat Eye retroreflector is several times brighter than the specified minimum. The measured value of the Cat Eye does not change with the change in observation angle.

Comparison of retroreflector light with cycle lamps

In the previously cited Johnson case, the plaintiff alleged that the tested Cat Eye retroreflectors were not as effective as a bike light since they did not give off any light to the observer. The plaintiff further alleged that the retroreflectors could not be seen even when captured directly in a beam of light from an automobile's head lamp.

In response to these implausible allegations, the light reflected by the Cat Eye retroreflector and three off-the-shelf bicycle head lamps were compared on a test track at various distances. The comparison, or test track illumination analysis, was conducted with a standard pair of car headlights on low beams. The photometer used to record the light was a Goshen Luna Pro on a tripod. All readings were converted to candlepower per foot-centimeter since that is the measurement unit employed by ASTM Standard ES09-91. The results of the illumination analysis are shown in Table II. The manner in which the analysis was conducted is shown in the Figure 2 schematic.

Results of the test track illumination analysis

The results of the test track illumination analysis yield interesting conclusions concerning the effectiveness of the CPSC retroreflector system. These conclusions are as follows:

1. When the CPSC retroreflector system is in the low beams of standard car head lights, the light reflected back to the motorist is equivalent to light reflected by standard bicycle head lights. Thus, the retroreflector system and bike headlights have, under these conditions, the same level of illumination.
2. When a cyclist is in the beam of standard motor vehicle head lights, the retroreflector system is moving. The movement further assists the motorist in identifying the object as a cyclist.
3. The test track analysis included tests to determine the amount of light reflected by the retroreflector system when the system is outside the beam of a car's headlight. Although no data points were developed suitable for publication, the test results revealed that the CPSC retroreflector system does not provide any feedback illumination when the cycle is outside the beam of a motor vehicle's headlight. The results further revealed that cycle head lamps outside the beam of a motor vehicle headlights will give the same degree of illumination as noted in Table II.

The above conclusions show that, when reconstructing bicycle accidents, the forensic engineer should carefully determine whether or not the cycle has CPSC reflectors. This is particularly true if the reconstruction involves the issue of night-time conspicuity. If the cycle is equipped with these reflectors and was in the beams of a motor vehicle's headlights, the cycle was clearly visible for a distance of 600 feet.

Using the Principles of Conspicuity to Reconstruct Bicycle Accidents

To evaluate the reaction times available to a cyclist and motorist(s), a forensic engineer must consider the conspicuity of the accident site at the time of the accident. As previously mentioned, conspicuity is, in bicycle accident reconstruction, the ability of those involved in the accident to see one another. This ability to see and be seen depends upon the prevalent ambient conditions, the colors worn by the cyclist, and the bicycle's colors and lighting (i.e., how conspicuous is the cyclist to others?). Optometrists give a more involved definition of conspicuity, but the reconstructionist needs to be concerned only with those issues that directly impact the determination of the accident causal factors. Thus, if reaction times are a major issue in the reconstruction, then only those conditions that directly impact the motorist's ability to see the cyclist should be factored into relevant equations.

In view of the foregoing, the purpose of this discussion of conspicuity is to define the major aspects of lighting and color that will directly impact the reconstructionist's determination of an accident's causal factors. These aspects of lighting and color were derived from actual field work and a review of the Engineering and conspicuity literature.

For organizational purposes, this analysis will focus on car-bicycle accident reconstruction. The same principles may, however, be applied in single cyclist accidents where the cyclist strikes an object or road abnormality. Again, this analysis focuses on the cyclist's ability to be seen and see his/her surroundings, including traffic.

Principles of Conspicuity in Nighttime Bicycle Accident Reconstruction

Nighttime collisions between cars and bicycles account for most car-bicycle accidents. Typical interviews after such accidents reveal that motorists simply did not see the cyclists. On the other hand, cyclists typically perceived that they could be seen by oncoming vehicles. Review of the literature and actual on-site calibrations reveal the following points that the reconstructionist should consider in these types of accidents:

- The current CPSC standards require that reflectors be placed on pedals, seat post, head set and wheels. Field verification has revealed that this reflector system does an excellent job at making a cyclist visible under nighttime conditions if the cycle is in the beams of a motor vehicle's headlights. For example, although Blomberg's subjects experienced some difficulty defining the retroreflector system feedback as coming from a cycle, this difficulty was not encountered when the cyclist was in the beam of car headlights.
- The Uniform Motor Vehicle Code does not currently mandate that a cyclist ride with a bike light a night. However, virtually all states now require cyclists who ride at night to insure that their bikes have lighting that is visible for 600 feet to the front and rear of the rider.

- Researchers have tested various lighting systems available on the consumer market. The results of these tests show that most available head lamps and tail lights cannot be seen by motorists at a distance of 600 feet. Despite this fact, cyclists believe that they are perceived by motorists at that distance. Only special lighting systems were found to provide illumination for the 600 feet distance.
- A cyclist's perception of objects at night is directly proportional to the lighting on the cycle. The definitive study in this area was done by David Sellers (see Appendix D: Bibliography). The results of this study were similar to those achieved in the aforementioned tests. Sellers found that the largest percentage of lighting systems were not adequate for perception of objects at night by the cyclist. The testing done under laboratory conditions revealed that lamp strengths failed to allow cyclists enough time to react to nighttime hazards. Using the results from this study, the reconstructionist should calculate the reaction time available to the cyclist based on the type of head lamp on the cycle. Underperforming head lamps will greatly reduce the reaction time available to the cyclist.
- When a motorist overtakes a cyclist at night, the type of rear bike lamp will have a substantial impact on the ability of the cyclist to be seen. I have analyzed and compared many of the available rear lamps with the lighting systems evaluated in the Sellers head lamp study and a Sellers study of reflectors.

Except for the Kearney system, all of the lighting systems failed to meet the Uniform Motor Vehicle Code standard of a minimum 600 feet visibility to the rear of the cycle. In several accident reconstructions in which I have been involved, the manufacturers stated (under deposition) that they were not even aware of the Uniform Motor Vehicle Code 600-foot requirement. Many cyclists have indicated during interviews that they perceived themselves to be visible at night with lighting systems when, in fact, they were not.

- Clothing for cyclists was found to have an impact on nighttime visibility. The lighter colors, as suspected, enhance conspicuity. However, the lightness of the clothing makes it harder for motorists to identify the object they are encountering. Field studies that I have conducted support the idea that motorists actually become confused when they encounter lighter clothing on a cyclist. Despite this possibility of confusion, the lighter clothing is preferable to dark clothing since it can be seen more easily.

- Fluorescent clothing is the most effective conspicuity aid a cyclist can use. The Blomberg studies show that perceived distances are increased from 150 feet to 560 feet with the use of fluorescent clothing. When factoring-in reaction times, the reconstructionist should allow for this increased perception distance in reaction time calculations. Independent studies that I have conducted reveal that the use of fluorescent paint on a cycle also greatly increases recognition by motorists.[3]

Field measurements that I have made showed perception distances which coincide very closely with the distances found in the Blomberg studies. According to these studies and measurements, fluorescent clothing increases the perception distance of a cyclist at night to almost 600 feet. I have also found that fluorescent clothing increases perception distances during the day from approximately 400 feet to 2,200 feet. Blomberg has done an excellent job of quantifying this increase in perception and has further shown that recognition of a pedestrian or cyclist increases from 290 feet to 720 feet with fluorescent clothing at night.

- Evaluation of nighttime bicycle accidents reveals that a cyclist without light or fluorescent clothing and an adequate lighting system is virtually unrecognizable to motorists. This is sometimes true even though the CPSC reflectors are in place. Interviews with drivers who have struck cyclists at night reveal that, with CPSC reflectors on the cycle, drivers are aware that something is in the road, but they are not always able to recognize that it is a cyclist. However, it must be stated again that, when a cycle with the CPSC retroreflector system is in the beams of a motor vehicle's headlights, the cyclist is recognizable. Under such conditions, the retroreflector system is as adequate as, if not more adequate than, most off the shelf, battery-operated lights.

In order to determine reaction times and accident causal factors, the reconstructionist should first determine the actual sight distances available to the cyclist and motorist. This determination must take into account the conspicuity available to the parties at the time of the accident. As noted at the beginning of this section, the reconstruction of car-bicycle accidents after dark is a function of the available reaction times, given the conspicuity at the site. Using the perception distances generated by the researchers in the bibliography, the visibility distances and available reaction times of the cyclist and motorist can be calculated. For example, when it is known what the cyclist was wearing, the Blomberg studies can be used to calculate sight distances. If the lighting system on the cycle is known, then the sight distances of the cyclist can be defined using the Sellers study. If time and money permit, it is best to determine actual reaction times in the field using the lighting conditions at the accident site, as well as the motor vehicle head lamps and distances involved in the accident. The work shown in Figure 2 is a good example of how accident parameters can be reproduced in the field.

Principles of Conspicuity in Daytime Bicycle Accident Reconstruction

Without the sight distance determinations needed for nighttime bicycle accident reconstructions, the reconstruction challenge for daytime bicycle accidents is one of visualization from clothing and safety equipment worn by the cyclist. The topography and roadway configurations are also more important during the day than they are at night. If the cyclist is not seen, or cannot see, other factors become negligible in determining the causes of an accident.

A review of the conspicuity literature and interviews with cyclists and motorists reveals the following factors for consideration:

- The type of clothing evaluated in the Blomberg studies is a factor during daytime accident reconstructions. Specifically, drab colored clothing, (i.e., gray, white, black, brown, light green, and light yellow) blends into the background as a motorist approaches a cyclist. This decrease in perception is illustrated by the fact that, although my analysis of reaction times showed that the motorists had several seconds to react to the presence of a cyclist, the drivers usually told interviewers that they simply did not see the cyclist. If one accepts the premise that no motorist deliberately runs into a cyclist, then the motorist's inability to see the cyclist should be taken at face value in these interviews.[4]

- The difference between the daytime and nighttime visualization of a cyclist is the type of clothing worn by the cyclist during the day. The more striking the colors, the easier the visualization for the motorist. As highway and traffic safety engineers have been aware for at least 20 years, fluorescent colors in yellow or orange make visualization more pronounced for motorists. Review of the conspicuity literature confirms this fact. When fluorescent colors are absent, a motorist's ability to see a cyclist fluorescent colors are made with the assumption that the cyclist was seen or should have can literally be present, reaction nonexistent. However, when time calculations should be been seen.

Unfortunately, as late as 1991, the manufacturers of cycling clothing and helmets often failed to take the conspicuity issue into account during the manufacture of their products. Helmet manufacturers have actually produced helmets with a camouflage coloring such as hunters might use. This virtually guarantees that the rider will blend with the background while in traffic. The manufacturers of cycling clothing have, however, recently started to take the issue of conspicuity into account. Yet, a significant percentage of the available cycling clothing, while being stylish and functional, still does not allow the cyclist to be seen by motorists. A bike jersey that may be acceptable in a European road race (where the course may be closed to traffic) may not be visible to motorists on a normal, busy road. The reconstructionist should allow for this fact when determining motorist reaction times. Lack of fluorescent or visually unusual clothing can make the cyclist essentially invisible during the daytime.

Conclusion

The effects of conspicuity on daytime and nighttime bicycle accident reconstructions are a function of the clothing worn by the cyclist. All conspicuity literature, field studies, and interviews with motorists reveal that this factor is very important in defining reaction times. Reconstructions of nighttime bicycle accidents must also take into account the efficiency and operational characteristics of the front and rear lighting systems on the cycle. In addition, the capacity of the motor vehicle lighting system must be determined when defining sight distances.

As noted in the various conspicuity studies, a motorist's perceptions are very important in his/her ability to visualize objects such as cyclists. Although Forensic Engineers routinely use statements of witnesses and parties when reconstructing an accident, it is extremely important that such statements be employed when conspicuity is a factor in the reconstruction. Perceptions of colors and distances can vary among individuals. Therefore, witness statements are very helpful in determining sight distances. If the situation needs to be quantified further, the services of an optometrist may be helpful.

Reaction time calculations are familiar to Forensic Engineers. The methodology of these calculations and the vectors for car-bicycle accidents are defined in Brown and Obenski's text on accident reconstruction. The physics of the pertinent vector analysis will not change with conspicuity information.

What will change is the available reaction time that the motorist or cyclist will have, based on the ability of the driver to see the cyclist and the ability of the cyclist to see his/her surroundings.

Endnotes

1. Collin Johnson vs. Derby Cycle Corp., et. al., Superior Court of New Jersey, Docket No. ESX-L-16063-89.
2. Blomberg, R.D.; Hale, A., and Preusser, D.F., "Experimental Evaluation of Alternative Conspicuity-Enhancement Techniques for Pedestrians and Bicyclists," JOURNAL OF SAFETY RESEARCH, 1986, Vol. 17, Pages 1-12. Later in this book the results of Blomberg's studies are examined in an effort to define how perception and reaction times are determined where conspicuity is a factor in nighttime accident reconstruction.
3. Ibid.
4. This may not always be a valid premise. I have interviewed motorists who opined that cyclists should not be on the roads and that anything that happens to them is their own fault for even being out there. Admittedly, these people are a very small segment of the drivers interviewed and they had other psychological disorders that could better be addressed by the criminal justice system.

The Utilization of the Principles of Conspicuity

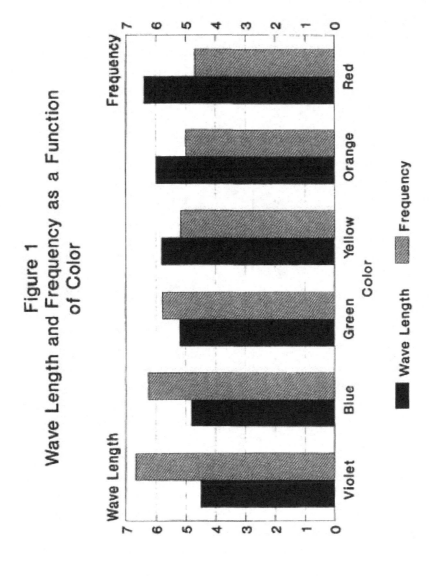

Figure 1
Wave Length and Frequency as a Function of Color

Wave Length = Meter•10-7
Frequency = Hertz•10+14

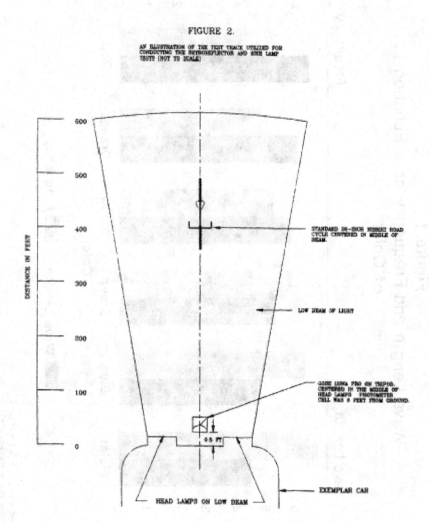

TABLE I

PHOTOMETRIC TEST RESULTS
CAT EYE REFLECTOR

Name of Device: Reflex Reflector
Type of Device: CAT EYE RR-290-WT
Name of Maker: CAT EYE CO., LTD.
Sample No.: 1447
Standard: C.P.S.C.

Date: August 17, 1992
Test Distance: 100 feet
Color: White
Unit: cp/ft-c

Observation Angle Deg.	Entrance Angle Deg.	Specified Minimum			Measured
		Red	Amber	White	
0.2	10U - V	4.50	11.25	18.0	43.48
	50L	1.50	3.75	6.0	9.41
	40L	1.75	4.37	7.0	11.76
	30L	2.00	5.00	8.0	14.72
	20L	2.25	5.62	9.0	29.17
	H - V	6.75	16.87	27.0	49.52
	20R	2.25	5.62	9.0	20.53
	30R	2.00	5.00	8.0	12.88
	40R	1.75	4.37	7.0	10.42
	50R	1.50	3.75	6.0	7.92
	10D - V	4.50	11.25	18.0	36.70
1.5	10U - V	0.05	0.12	0.20	0.698
	50L	0.03	0.07	0.12	0.542
	40L	0.03	0.07	0.12	0.634
	30L	0.03	0.07	0.12	0.537
	20L	0.03	0.07	0.12	1.524
	H - V	0.07	0.17	0.28	1.309
	20R	0.03	0.07	0.12	0.627
	30R	0.03	0.07	0.12	0.499
	40R	0.03	0.07	0.12	0.595
	50R	0.03	0.07	0.12	0.470
	10D - V	0.05	0.12	0.20	1.160

TABLE II

A COMPARISON OF THE CAT EYE RETROREFLECTOR WITH BICYCLE HEAD LAMPS AT VARIOUS DISTANCES
(Results are Expressed as cp/ft-c)

Distance (ft)	Retroreflector	Cat Eye I	Cat Eye II	Performance
50	80	81	84	92
100	45	45	42	47
200	41	42	44	40
300	39	41	39	42
400	35	38	34	39
500	28	33	30	32
600	8	11	10	9

Analysis was conduced in August, 1993
at Research Triangle Park, North Carolina

Cat Eye I, Cat Eye II, and Performance = off the shelf battery operated bicycle head lamps.

See Figure 2 for the experimental set up. It should be emphasized that the retroreflector system gave comparable results to the standard bicycle head lamps within the motor vehicle cone of light.

Chapter 10

Calibration of Nighttime Bicycle Accident Sites Using the Principles of Conspicuity and the Grid System

Introduction

As previously discussed, the reconstruction of nighttime bicycle accidents involves the utilization of conspicuity principles. Conspicuity, for purposes of bicycle accident reconstruction, is defined as "the ability of the cyclist to see or be seen." The purpose of this analysis is to develop, for the Attorney or Forensic Engineer, the necessary methodology to calibrate an accident site when the accident reconstruction involves issues of conspicuity. The engineering terms will be identified and a case study used to illustrate the application of conspicuity principles to a bicycle accident reconstruction.

Conspicuity in Bicycle Accident Reconstruction

The ability to see or be seen is a function of those areas of light that can affect the sensor (retina) of the human eye. The human eye is most sensitive to light at a wavelength of 555 lambda (k). The wavelength is related to the speed of light, C, by:

$$C = f k$$

where;

$C = 1.86 \times 10^5$ miles/second

$f = 3.8 \times 10^{14}$ to 7.9×10^{14} Hertz

The Forensic Engineer is interested in measuring light at the accident site in terms of lux. Lux is a measure of illumination and is a function of lumens. The latter term is the amount of light reflected from the pavement. The scientific definition of lumen is "the luminous flux emitted in a solid angle of one steradian by a uniform point of source of intensity of one candela." Another definition of a lumen is 1/680 watt of yellow-green light of wavelength 550 m at 100 percent intensity.

A more practical method of applying these scientific definitions entails using the same methods employed by professional photographers. The forensic engineer essentially calibrates the site by using a photometer that can accurately define visual light in terms of lux or lumens. The meter also should convert to foot-candles in case the forensic engineer wants to compare that parameter.

VISIBILITY - The visibility of a cyclist is, in essence, its conspicuity. We will assume that the size of the cyclist and the cycle is a constant for purposes of defining conspicuity and visibility. The true definition of conspicuity then becomes a study in contrast.

CONTRAST - Is the difference in the luminance of the object of interest and the luminance of the background that the object is seen against.

LUMINANCE - This parameter is the quantitative method for measuring light from a source. Luminance is different than brightness. Luminance is measured by instrumentation while brightness is the sensation of what we perceive. Brightness depends on the adaptation of the eye and on the relative brightness of areas surrounding the small spot whose brightness we are judging. The moon looks bright against the dark sky and dim when seen against the daytime sky. Therefore, the moon's brightness seems to vary, but its measured luminance is a constant value.

Luminance is measured with an instrument that has an optical system and internal masks which frame the area to be measured. The basic measurement then becomes candelas per square meter (Cd/m2) or foot lamberts per square meter (FL/m2). Pavement luminance at a point is calculated by the following formula:

$$L = \frac{I \times r}{H^2 \times 10,000}$$

where:

L is luminance in Cd/m^2
I is the light intensity in candelas
r is the reduced coefficient of pavement reflectance
H is the height of the luminaire (i.e., street light) above the calculation plane in meters.

Generally, luminance is used to define the amount of light being reflected from a surface or pavement. Luminance values are used in designing lighting systems. Illumination is of much greater interest to the forensic engineer. A more simplistic view of the relationship can be given as:

Luminance = illuminance x reflectance

ILLUMINATION - In pedestrian or bicycle accident reconstruction, is the amount of light being emitted from a light source. It is designated by the symbol lux or Lx and is the illuminance of one lumen per meter squared or 1 lm/m^2. The relationship of interest in bicycle accident reconstruction is:

$$1 \text{ lux} = Lx = 1 \text{ lm/m}^2 = cd/m^2$$

For the reconstructionist, this relationship is very powerful at the accident site. To measure ground level accident site light data, one can employ a well-calibrated light meter such as the Gossen Luna-Pro, or equivalent, and a grid can then be used to measure and record accident site level values of light at each node with the light meter. The values can then be compared to previously determined minimum values of lumination to define the ability of the parties to the accident to visualize one another.

GRIDDING THE SITE - From the probable point of impact (POI), for a distance of at least 600 feet prior to the POI, a grid should be prepared with each node a maximum ten feet apart. Ideally, it would be appropriate to have an infinite number of nodes. However, time and cost become a tremendously limiting factor in laying out the grid. Light readings should be taken with the meter facing the direction of the eye of the motor vehicle driver. The light meter should "see" the same illumination as the human eye.

In bicycle and pedestrian accident reconstruction, the analyst is interested in the illumination. For purposes of defining the adequacy of the ambient conditions, the amount of light reflected from the pavement is lumination and generally is not of interest since all values are in lux on the grid and are assumed to be illumination.

When the area is defined in terms of an illumination value at each node, it is possible to connect lux values of the same magnitude by a line. These interconnecting lines from node to node are called isohyets and enable the engineer to define the illumination pattern at the accident site. The engineer can then define the capability of the cyclist or pedestrian to be "seen" by the motorist.

Reconstruction of a Night Accident: A Case Study

A cyclist dressed in a black uniform was riding a cycle on the right side of the road at 11:30 p.m. The cyclist was struck from the rear by a hit-and-run driver. An investigation was conducted to determine the causes of the accident using the following additional data:

- Street lights were present on the right hand side, approximately 160 feet apart.
- The road had very little traffic flow at that time of day.
- A golf course sprinkler system was operating in sequence approximately 20 feet to the right of the cyclist.

The steps in the investigation consisted of gridding the area in 10 foot nodes 600 feet prior to the POI and 100 feet after the POI. The POI was determined by using a classic trajectory analysis. The illumination values revealed that an area under the street lamps gave a 15 lux reading in a circle 8.5 feet in diameter. The isohyets showed a 1.4 lux reading in the areas not superintended by the street lamps. The golf course sprinklers discharged in a semicircle at a rate of 75 gpm in a direction away from the roadway. However, winds of 10-15 mph were recorded under the meteorological conditions at the time making it probable that spray from the sprinklers was blowing out into the road. The gearing on the cycle gave a probable speed of 12 mph at 50 rpm pedal cadence. An analysis of the reflectivity of the clothing revealed that no incidental light was reflected back at an angle of 30° one foot from the clothing.

Engineering Analysis of the Facts of the Accident

Certain facts became apparent from examining the grid of the accident site:

- Due to the cyclist's clothing, lack of lighting and reflection on the cycle, and the background values of 1.4 lux, the rider was virtually invisible unless under the street lamp.
- Due to a "pool" of light under the street lamp of 15 lux, the cyclist could be seen at a velocity of 17.64 ft/sec for 0.48 seconds as the rider passed under the street lamp.
- The water spraying into the road could have caused the "invisible" cyclist to swerve into the lane of traffic. There was no water on the cyclist five minutes after the accident, so the water spray was an unlikely cause of the accident.

After gridding the site and developing the isohyet, it was apparent that the cyclist was only visible during the 0.48 seconds that he passed under the street lights. No municipal or commercial lighting design could have provided adequate lighting for the cyclist to be seen. The fact that the cyclist was dressed in black and had no reflectors or lighting on his bicycle was the cause of the accident.

Principles of Conspicuity in Bicycle and Pedestrian Accident Reconstruction

The Department of Transportation guidelines are adequate for the design of municipal roadway lighting systems. These systems provide the spatial relationships between mounted lights as a function of pavement luminance. The design of lighting systems must assume a linear distance between the lights due to pure cost effectiveness. The cyclist and pedestrian must assume a degree of the responsibility to be seen. The engineering community has painstakingly worked out a standard for cyclists and pedestrians to be seen.

A light should be in place on the cycle such that the rider can be seen 600 feet to the rear at night. Photometric analysis of the lights on the market through 1991 revealed that all cycle lamps on the market met this minimum requirement for the rear at night. Prior to 1991, most cycling lamps tested did not meet this criteria. During the day, fluorescent lime green, yellow or orange allows the cyclist to be seen at approximately 600 feet, as well. The basic principles for reconstruction of cycling or pedestrian accidents are as follows:

1. During nighttime reconstruction, grid off the site in 10 foot (or equivalent) nodes. Measure the illumination at the initial points leading up to the impact zone in terms of lux. Foot-candles can be used, but field investigations tend to be more effective when the data is collected in lux units. Collect the data with the light meter taking the position of the driver's eye.
2. Compare the illumination data at the scene with known data on the minimum amount of illumination needed to allow the cyclist to be "seen." This known data is contained in Department of Transportation documents and in research conducted by optometrists. This information also is provided in summary form in the following list.
3. Basic freeway and roadway lighting using approved designs yields illumination data of 2 lux at the area of lowest effective output of the lighting system. Lighting at the brightness area should be 6 to 9 lux. Both readings assume average maintained horizontal illuminance.
4. In areas where background illumination is 2 lux (such as rural areas), basic studies in night perception yield virtual invisibility if a pedestrian or cyclist is dressed in dark clothing and does not have reflecting material or a light. Even with vehicular traffic present, and adequate roadway illumination, cyclist and pedestrian percep-

tion under these conditions is about 75 feet in a surprise situation. At even 35 mph or 51.5 feet/sec, a motorist has only 1.46 seconds to perceive the object and avoid it. This leaves the motorist with an almost impossible maneuver, since it takes almost three seconds in a surprise situation to insure complete avoidance.

5. In the same rural or dark areas of 2 lux illumination, if the cyclist or pedestrian has a reflecting vest or equivalent, detection distances vary from 1,200 to 2,200 feet. Recognition distances vary from 600 to 700 feet. Again, passing motor vehicles do not seem to affect the physiological ability of the eye to perceive. The use of a cycle light or flashlight capable of being seen 600 feet to the rear was found to insure recognition. (It should be noted that several states require cyclists to have a light which meets that criteria while riding.)

6. In suburban areas, where background lighting is 2 lux and artificial light provided by light poles (luminaries) placed 130 feet apart and staggered centrally on opposite sides of the roadway is such that the illuminated areas were 6 to 9 lux, recognition distances are shorter. For a reflected cyclist or pedestrian, these distances vary from 260 to 325 feet. No detection distances have been found in the engineering literature. The shorter recognition distances are due to the "lighting noise" created by traffic and peripheral effects. The added stimuli in the urban setting makes observance of a reflected cyclist more difficult, thus decreasing the recognition distance.

7. Reflectors on a bicycle are required by the Consumer Product Safety Commission (CPSC). Reflectors are required on both sides of the pedals, on the frame directly behind the seat, on the headset and on the wheel spokes. Studies have shown that, when approached from behind by a motor vehicle, in a rural area, a cycle with reflectors can be detected at 600 feet.

Recognition, however, is dependent upon the type of clothing worn by the cyclist. The previously defined recognition distances will prevail. When approached from the side, the rotating motion of the reflectors on the wheels yields detection and recognition at 600 feet for the cyclist.

Conclusion

The ability of cyclists and pedestrians to be seen is really a function of the reflective clothing and lighting on the cycle and individual. It is impossible from an engineering viewpoint to design a roadway that will enable a cyclist to be recognized and detected at safe distances. In urban settings, where ambient lighting illumination approaches 10 lux, the background lighting present actually decreases recognition and detection distances.

When reconstructing a bicycle or pedestrian accident where lighting is a factor, the forensic engineer should grid the accident site. The illumination in terms of lux, or equivalent, should be determined at each node. The values should be obtained from the view point of the motorist. The gridded area can then be compared to the minimum light required to be seen. It should be emphasized that only under light poles does enough illumination occur for detection and recognition of the cyclist or pedestrian. Any other detection and recognition is dependent on the lighting and reflection on the cyclist or pedestrian.

Chapter 11

A Determination of the Reaction Times Available to a Motor Vehicle Driver Overtaking a Cyclist at Night

Introduction

A cyclist struck from the rear by a motor vehicle at night is a predominant feature of fatal accident statistics. In an attempt to change this disturbing fact, many studies have been conducted to determine the visibility of various colors at night. These studies have found that the use of light-colored clothing can substantially increase the distance at which a cyclist can be seen. Also, the use of a light that is visible 600 feet in the rear can greatly increase the ability of a motorist to see and react to a cyclist. Several states require rear lighting on cycles that is visible for 600 feet to the rear. The purpose of this analysis is to define the reaction time available to a motorist under different scenarios. In addition, the methodology for determining reaction times will be defined and the importance of the 600-foot rule will be emphasized.

Case I: Assume No Reflecting Devices on the Cycle, Dark Clothing on the Rider.

In this scenario, the car overtakes the cyclist at 30 mph at night. The cyclist has no reflective clothing nor any lighting on the cycle. Visibility is not affected by ambient conditions. Given that the driver has 1.0 second to perceive the cyclist and 1.0 second to react to the surprise situation, the determination of travel distance prior to braking can be calculated as follows:

$$(30 \text{ mph}) (1.47 \text{ ft/sec/mph}) = 44.1 \text{ ft/sec} = \text{Velocity}.$$

Therefore,

$$(1.0 \text{ set}) (44.1 \text{ ft/sec}) + (1.0 \text{ set}) (44.1 \text{ ft/sec}) = 88.2 \text{ ft of travel distance prior to braking}$$

In this instance, braking distance is ignored since it will vary between motor vehicles. Also, braking distance calculations will not change the fact that the cyclist is at risk in the analysis since the detection distance of the cyclist will be less than the braking distance.

Detection distance is determined by Blomberg[1] as 70 feet with no reflection on the cycle or the rider. The cyclist will continue to travel while the motorist attempts to react to the surprise stimulus. Since the cyclist is assumed to travel at 10 MPH, the distance the cyclist will travel is defined as:

$$\text{Distance} = (10 \text{ mph}) (1.47 \text{ ft/sec/mph}) (2.0 \text{ set}) = 29.4 \text{ ft.}$$

The Zone of Containment (Z_C) of the car is defined as "the distance in front of the car that the vehicle cannot avoid in a surprise reaction situation." It will be defined, in this case, as the second that occurs after the first surprise second, where avoidance of the cyclist might occur. Calculation of the zone is as follows:

$$Z_c = (1.0 \text{ set}) (44.1 \text{ ft/sec}) = 44.1 \text{ ft.}$$

In Case 1, the calculated results are:

- Speed of Car = 30 mph
- Distance Car Travels Prior to Braking = 88.2 ft.
- Detection Distance of Cyclist = 70.0 ft.
- Zone of Containment of Car = 44.1 ft.
- Distance Cycle Travels While in the
- Zone of Containment = 14.7 ft.

The car will travel the full distance of the Zone of Containment after striking the cyclist (44.1 Feet - 14.7 Feet = 29.4 Feet). The primary reason for this is that the detection distance of the cyclist is much less than the distance the car travels prior to braking. As a result, the motorist cannot detect the cyclist soon enough to prevent the cyclist from entering the Zone of Containment.

Lack of reflective clothing and/or lack of a rear light which can be seen from 600 feet to the rear lowers the detection distance to the point that the cyclist is almost certain to be struck by the motor vehicle. Also, an increase in the motor vehicle's speed will decrease the detection distance, as the Blomberg studies point out[2]. For comparison purposes, a case analysis generated at 55 mph is set forth below.

Case II: Car Approaches Cyclist From Behind at 55 mph.; The Cyclist Has No Reflective Clothing.

In Case II, the results are as follows:
- Speed of Car = 55 mph
- Distance Car Travels Prior to Braking = 161.8 ft.
- Detection Distance of Cyclist = 70 ft.

•Zone of Containment of the Car = 80.9 ft.

As in Case I, the car will travel the full distance of the Zone of Containment (Z) after striking the cyclist (80.9 ft. - 14.7 ft. = 66.2 feet). In both Case I and Case II, the distance the cyclist travels in the Zone of Containment will be somewhat less than 14.7 feet, depending on the speed of the car at the beginning of the surprise reaction sequence.

Conclusion

A large number of accidents involving motor vehicles overtaking cycles from the rear occur at night. As Blomberg and others have pointed out, the available detection distance does not afford the driver an opportunity for timely reaction to the nearly invisible cyclist. Based on this analysis of the available detection distances, the following conclusions can be drawn:

1. The extensive amount of work that the Department of Transportation (DOT) has generated in its effort to define a standard for nighttime cycling has resulted in the 600-foot rule. The rule provides that a cyclist must have a rear light on the cycle that is visible 600 feet to the rear of the cyclist. As illustrated by the two cases detailed above, the 600 feet of detection distance will easily enable a motorist to see and avoid the cyclist.
2. The Consumer Product Safety Commission (CPSC) requires that all cycles be equipped with reflectors at specified locations on the cycle. These reflectors increase the perception distance of the cycle. However, the recognition distance of the cycle will not necessarily increase. My bicycle accident reconstruction work has revealed that cycles equipped with CPSC required reflectors are perceived by motorists, but not always recognized as bicycles. Motorists frequently stated that they didn't identify the reflective object as a cyclist until it was too late to avoid the collision. Perception distances varied (i.e., depending upon the methodology used to measure perception), but were usually much less than the 600 feet provided by rear lamps.
3. There is a difference in contrast between the reflectivity of a cyclist's clothing and background ambient light. That is, a motorist can discern at night that some type of object may be ahead, but cannot tell what the object is if no lights are on the cycle. This appears to be true, although the cyclist thinks he or she is visible. Similarly,

reflectors on the cycle will increase perception distance, but the contrast in reflectivity between ambient lighting conditions and the clothing worn by the cyclist is not sufficient for the cyclist to be recognized by the motor vehicle.

Endnotes
1. Richard D. Blomberg, Allen Hale and David F. Preusser, "Experimental Evaluation of Alternative Conspicuity - Enhancement Techniques for Pedestrian and Bicyclists," Journal of Safety Research, Vol. 17, pp. 1-12, 1986.
2. Ibid.

Chapter 12

The Derivation of a Formula for Determining the Speed of a Bicycle Rider Down an Incline

---

Assume a bike rider down an incline:

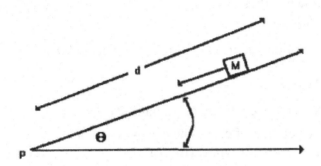

Given:

    Distance traveled down incline, d
    Angle of incline, Θ
    Mass of bike frame, M = 18 lbs.
    Mass of wheel, M_2 = 1 lb/wheel
    Mass of rider, M_3 = 100 lbs.

Find:

    Velocity at point P, bottom of incline = V_P

Assume:

    Initial velocity, $V_0 = 0$
    No resistance

85

Nomenclature

M_r = mass of rider
M_f = mass of frame
M_w = mass of one wheel (including tire)
M = total mass = $m_r + m_f + 2m_w$
r_w = radius of wheel (to outer edge of tire)
C = center of mass of one wheel (at axle)
G = center of mass of entire system (bike and rider)
d = distance down incline
Θ = angle of incline
g = local acceleration due to gravity = 32.2 ft/sec^2 = 9.81 m/s^2

Kinematic Terms

x = angular velocity of each wheel (radian/set)
a = angular acceleration of each wheel = dw/dt (rad/sec^2)
V_c = velocity of center of wheel (ft/sec)
V_G = velocity of center of mass of entire system (ft/sec)
A_c = acceleration of center of wheel
A_G = acceleration of center of mass of entire system (ft/sec^2)

Note: $V_c = V_G$ and $A_c = A_G$ for G fixed in system.
For no slippage of the wheel on the track:

$$V_G = V_C = r_w \omega$$

$$A_G = A_C = \frac{dV_c}{dt} = r_w \alpha$$

Preliminary Calculation of the Mass Movement of Inertia of a Wheel

We need to express the movement of inertia (a measure of the resistance to change in the rotational motion) of each wheel as follows:

$$I_c = \int r^2 dm$$

Integrated over the entire mass of the wheel.

To estimate I_c theoretically, we need to know the distribution of mass in the wheel (hub, spokes, rim and tire). If the entire mass is concentrated at the outer radius, r_w, then $I_c = \int r_w^2 dm = r_w^2 m_w$.

Let $I_c = k^2 r_w^2 m_w$ where k is a constant that is less than but approaching 1, since most of the mass of the wheel is near its outer radius.

Review of Force-Mass Acceleration Approach

Newton's 2nd Law of Motion written in the "x" direction:

$$\sum F_x = m a_{Gx} ;$$

$\sum F_x$ = sum of the x components of all forces acting on the body of mass m

a_{Gx} = x-component of the acceleration of the center of mass of the body.

For a symmetrical rigid body in plane motion:

$$\sum M_c = I_c \alpha ;$$

$\sum M_c$ = sum of the moments about an axis through the center of mass of the body. The axis is perpendicular to the plane of motion.

I_c = mass moment of inertia with respect to the axis through C.

α = angular acceleration of the body.

The problem is to determine if the wheels are negligible to the mass of the rider.

Solving by the F = ma approach, neglecting the resistance of air:

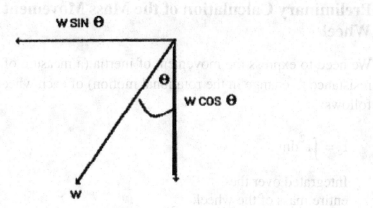

$V_0 = 0$ since riders are not moving at the starting gate
$V_P = V$ or velocity at the bottom of the ramp
$F_1 + F_2$ = tangential components of track on tires
$N_1 + N_2$ = normal components of force of track on tires
$A_G = a$ = acceleration of bike and rider
$W \sin \Theta$ = vector components of rider mass
$w \cos \Theta$

Substituting the weight components into the F = ma equation

$$(1) \quad mg \sin 0 \; F_1 - F_2 = ma$$

If the wheel has the following components:

Therefore:

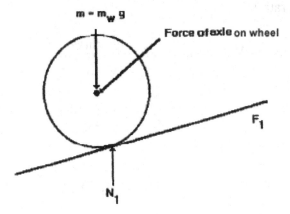

$$\sum M_c = I_c \alpha$$

$F_1 r_w$ = moment bearing friction = $I_c \alpha$ where the friction from the bearings is negligible

$I_c = k^2 r_w^2 m_w$ Derived previously, and

$$\alpha = \frac{a}{r_w}$$

Thus:

$$(2)\ F_1 r_w = k^2 r_w^2 m_w \left(\frac{a}{r_w}\right) = F_1 = k^2 m_w a$$

This is the freebody diagram of one wheel.

$$(3)\ F_2\ k^2 m_w a = \text{freebody diagram of second wheel}$$

Substituting (2) and (3) into equation Number (1)

$$mg \sin \Theta = 2k^2 m_w a = ma$$

$$a = \frac{M_g \sin \Theta}{m + 2k^2 m_w} = \text{constant if } F_1 = F_2 = \text{constant}$$

(It is assumed that both wheels are on the ground.)

For constant acceleration:

$$V^2 = V_0^2 + 2ad = 0^2 + 2\frac{(mg\sin\Theta)(d)}{m + 2k^2 m_w}$$

$$V_p = V = \left[2\frac{(mg\sin\Theta)(d)}{m + 2k^2 m_w}\right]^{0.5}$$

$$d\sin\Theta = h \text{ where } h = \text{height of start}$$

This is the equation of a bike rider going down an incline using the F = ma approach.

Work Energy Approach

$$\text{Work} = \text{change in Kinetic Energy } (U = \Delta T)$$

or

$$T_0 = V_{go} + \text{Energy Lost} = T_p + V_{gp}$$

Where:

P = point P (end)
T = kinetic energy
V_g = gravitational potential energy

The equation then becomes:

$$0 + mgh - 0 = \frac{1}{2}mV_G^2 + 2\left[\frac{1}{2}I_c\omega^2\right]$$

Where:

$\frac{1}{2}mV_G^2$ is translational kinetic energy and

$2\left[\frac{1}{2}I_c\omega^2\right]$ is rotational kinetic energy of each wheel

Therefore:

$$mgh = \frac{1}{2}mV_G^2 + k^2 r_w^2 m_w \omega^2$$

Substitute $h = d \sin\Theta$, $V_G = V$,

$$mgd \sin\Theta = \frac{1}{2}mV^2 + k^2 m_w V^2 = V^2 \left[\frac{m + 2k^2 m_w}{2}\right]$$

$$V_p = V = \left[\frac{2mgd \sin\Theta}{m + 2k^2 m_w}\right]^{0.5}$$

If the mass of the wheels is assumed to be negligible, then:

$$V_p = \left[\frac{2gd \sin\Theta}{m}\right]^{0.5} = [2gd \sin\Theta]^{0.5} = [2gh]^{0.5}$$

The above derivation essentially achieves the theoretical free fall equation. I have used this equation in a number of reconstructions to determine conceptual speeds at accident sites and found it to be fairly accurate.

One individual has criticized the free fall equation on the grounds that it fails to consider wind resistance. (It should be noted that this individual has, to date, failed to produce his own equations.) The equation does omit consideration of wind resistance. This omission is due to the negligible impact that wind resistance has on achieved results. I have, however, set forth below for the benefit of these critics the calculation for utilizing wind resistance that is cited in most engineering literature. [1,2]

Drag Force (lb F) = 0.00256 x C_D x front area ft^2 x [speed (mph)]^2

Where:

C_D = Coefficient of Drag = 0.9 for a bicycle[1]

Since the free fall equation, with or without consideration of wind resistance, can only be used to obtain conceptual speeds, field tests should be conducted where practicable to determine the cyclist's actual speed.

A Field Study

A series of test hills on a 100 foot long asphalt test surface were utilized to compare the free fall equation with actual cycling speeds. A 150 lb. cyclist on a Raleigh racing road bike frame with clincher tires started down the hills from a stopped position. As the cyclist descended the hills, his speed was recorded.

The results of the comparative study are noted in Figure 1. Note that the theoretical free fall speed was calculated and plotted on Y_1. The actual speed measured was recorded on Y_2.

Conclusion

The engineering literature and the field tests noted in Figure 1 indicate the same conclusion with respect to the determination of a cyclist's speed down an incline. Namely, there are too many variables, regardless of the formula used, to predict the cyclist's speed with a high degree of certainty. In light of this fact, field tests should, as noted above, be conducted to determine the cyclist's speed down an incline. Initial field verification of Burns' and Sullivan's work indicate their coefficients, graphs, and formulas to be the state-of-the-art in determining speed of a cyclist under all conditions.

Endnotes

1. See, for example, Whitt, F. and Wilson, D. "Bicycling Science." Second Edition. Cambridge, MA: MIT Press, p. 92, 1989. In addition, the relationship between drag force and speed is explored in Burke, E. Editor. "Science of Cycling." Champaign, IL: Human Kinetics Publishers, Inc., pp. 92, 123-136, 1986.
2. Bums, Steven P. and Sullivan, John P., "Aerodynamics of the Rider - Bicycle Combination," *Cycling Science,* Vol. 7 , No. 1 , Fall 1995, p p . 8 , 9 , 10, 2 2 , 2 3 .

FIGURE 1: VELOCITY OF A 150 POUND RIDER AT VARIOUS SLOPES COASTING FROM A STOP FOR A DISTANCE OF 100 FEET

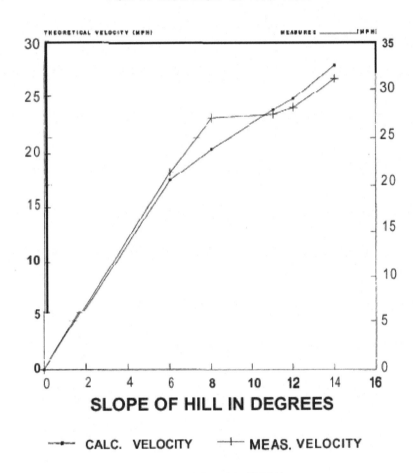

DERIVED FORMULA USED FOR THEORETICAL SPEED

Chapter 13

Basic Engineering and Physics Applicable to Bicycle Accident Reconstruction

Introduction

The purpose of this chapter is to present the basic equations that are used in reconstructing bicycle accidents. There are many excellent texts available that present basic Engineering and Physics for accident reconstruction. However, the equations presented in the texts are often not specific to the reconstruction of cycle accidents. In contrast, the equations contained in this chapter have been field-tested and are unique to bicycle accident reconstruction.

Applying generalized equations to the reconstruction of bicycle accidents can be risky, since the physics of the cycle are unique. I have, therefore, attempted to clarify the appropriate formulas that can be used in cycle accident field investigations.

Equations

Determination of the speed of a cycle while descending a hill:

To measure velocity in the field from a stop:

$$V = aft$$

Where:

$a = 32.2 \text{ ft/sec}^2$

f = acceleration factor (dimensionless)

t = time of travel from $V = 0$ to measured point

To measure velocity in the field where the cycle has an initial velocity at time zero:

$$t = \frac{V_0 - V_f}{af}$$

Where:

V_0 = initial velocity (ft/sec)

V_f = final velocity (ft/sec)

a = gravitational constant 32.2 ft/sec^2

Deriving the final equation:

$$V_f = [V^2_0 - 64.4fd]^{0.5}$$

Where:

d = distance in feet

Determination of the acceleration factor f is quite important in bicycle accident reconstruction. To make this determination, the three equations set forth below should be employed in the field.

$$f = \frac{V}{at}, \quad f = \frac{1.466s}{at}, \quad f = \frac{0.0455s}{at}$$

Where:

f = acceleration factor (dimensionless)
t = time in seconds
v = velocity of feet per second
s = speed in miles per hour
a = gravitational constant, 32.2 ft/sec^2

Values of f in the field vary as a function of slope. For example, a cyclist travelling 28 mph on a 5% slope will yield an f value of approximately .04.

I have derived the following equation for determining the speed of a cyclist down an incline or slope. This derivation is the result of input from other engineers and field verification. The model derivation for this equation is set forth in the preceding chapter.

$$V = [2gd \sin \Theta]^{0.5}$$

Where:

g = gravitational constant (32.2 ft/sec2)

d = distance on the slope in feet

Θ = degree of slope

This can be condensed to:
$$V = [2gh]^{0.5}$$

Where:

h = the height of the cyclist above, the ending elevation of the slope.

This equation tends to create problems if the cyclist's distance of travel is over 50 feet. At that point, wind resistance becomes a factor. The recommended equation then becomes:
$$V = [2gd \sin \Theta]^{0.5}$$

Note that, due to variances in topography and wind resistance, the above equations should be tested in the field whenever possible. Note further that these equations are not intended for use in obtaining final speeds. Rather, they should be used to achieve initial speed estimates at the beginning of the investigation.

Most police accident reports and site surveys give slope as a percentage value. Figure 1 can be used to convert such values to angular degrees. The latter is necessary to substitute for Θ in the formula.

It should be noted that most reconstructionists overestimate cycling speed. Some facts that are helpful in avoiding this problem are as follows:
- A world class time trialer can make the cycle average between 25 mph and 30 mph. This is someone who trains 500-700 miles per week.
- Typical speeds of the casual cyclist are about 10-12 mph on level ground.

Determination of Trajectory

It is often necessary to define the trajectory that a cyclist will travel after impact (or following projection from the bike due to any number of reasons). The appropriate equations are as follows:

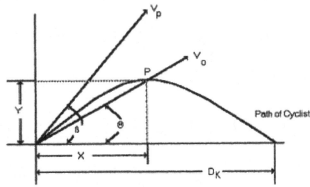

Fig. 1 Spatial Relationship on the Trajectory of a Cyclist

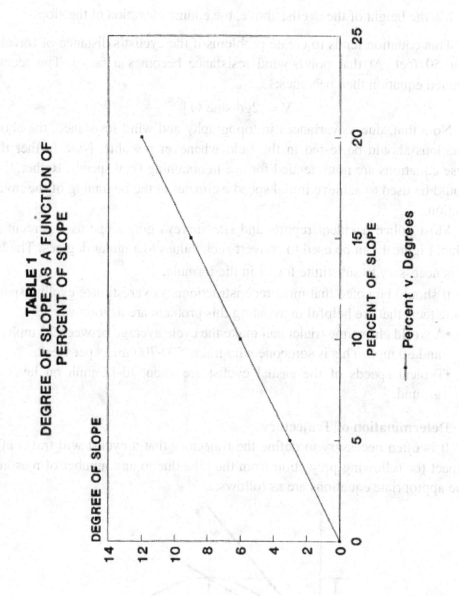

The trajectory of the cyclist has both horizontal and vertical components. Assume the original velocity of the cyclist at point zero is v_O feet/second making an angle of β degrees with the horizontal. Therefore,

$$v_x = v_0 \cos\beta \text{ feet/second}$$

The horizontal distance (x) traveled in t seconds at the original velocity of v_0 becomes,

$$x = v_0 t \cos\beta \text{ feet}$$

where:

a downward uniform acceleration of (a = 32.2 feet/second^2) is assumed.

The vertical component of velocity (v_Y) at the end of t seconds of a particle having the same initial velocity of v_0 can be seen as:

$$v_Y = v_0 \sin\beta - at \text{ feet/second}$$

The vertical distance (Y) traveled in t seconds uses the same components in following relationship:

$$Y = v_0 t \sin\beta - at^2 \text{ feet}$$

The above relationships can be derived to reveal points on the trajectory path that are of interest in defining the cyclist's airborne travel. All relationships assume that the initial velocity is V_0 feet/second at β degrees from the horizontal with the normal downward acceleration of: a feet/second^2. These additional formulas are:

Time (t_v) to reach the highest point on a path;

$$t_v = \frac{v_0 \sin\beta}{a} \text{ seconds}$$

Vertical distance (d_v) from the horizontal to the highest point of the path;

$$d_v = \frac{v_0^2 \sin^2\beta}{2a} \text{ feet}$$

Velocity (v) at the end of t seconds;

$$v = [v_x^2 + v_y^2]^{0.5} = [V_0^2 - (2V_0 \, at \sin\beta + a^2 t^2)]^{0.5} \text{ ft/sec}$$

Time (t_h) to reach the same horizontal point as at the start of the trajectory;

$$t_h = \frac{2v_0 \sin\beta}{a} \text{ seconds}$$

This equation is used when the cyclist is projected at a point above the coordinates X = 0, Y = 0, but eventually comes to rest on the horizontal at that point.

The horizontal distance (d_h) traveled when the cyclist is projected and returns to the same point on the horizontal;

$$d_h = \frac{v_0^2 \sin 2\beta}{a} \text{ feet}$$

The time (t) to reach any point (P) on the cyclist's trajectory if the line through point p and the starting point makes Θ degrees with the horizontal (see Figure 1);

$$t = \frac{2v_0 \sin(\beta - \Theta)}{a\cos\Theta} \text{ seconds}$$

All equations assume that the cyclist's center of mass is point P on the trajectory. The center of mass of the cyclist's body is the same as the center of gravity. The center of mass is that point through which the resultant of the weights of all the components pass, whatever the position of the cyclist's body. Simply put, it is a point in the center of the cyclist's body approximately 1/2 of the distance from a line through the top of the pelvic girdle and a line through the bottom of the last pair of ribs on the rib cage.

In calculating trajectories, the mass of the cyclist and its direction are of primary importance. The weight of the bike is negligible compared to the weight of the cyclist in trajectory determinations.

Shear Force on Forged-Rolled Steel and the Application of These Forces to Bicycle Accident Reconstruction

When a cycle is struck a glancing blow by a motor vehicle, it is possible to draw some conclusions about the motor vehicle's speed from an inspection of the cycle. Of particular interest is the shearing of any bolts, such as axle bolts, pedal bolts, seat post bolts, and head-set bolts. Shear, in its application to bicycle accident reconstruction, is defined as "that point where the intensity of stress caused by impact results in a tearing of the metal." Shear values are

Basic Engineering and Physics

unique to the types of steel as follows:

Table 2
Shear Versus the Type of Material

Material	Shear (PSI)
Steel, Forged-rolled:	
C, 0.10 - 0.20	48,000
C, 0.20 - 0.30	53,000
C, 0.30 - 0.40	56,000
C, 0.60 - 0.80	75,000
Nickel:	92,000

The magnitude of the vehicle speed can be determined if shear is noted. If we assume a 150 lbs. mass cyclist that is sideswiped by a 3,500 lbs. mass automobile resulting in the shearing of a rear axle bolt with a C, 0.15 steel, forged rolled component then;

The relationship between velocity and shear can be expressed as:

$$v_1^2 = \frac{\varsigma_s \pi d^2 (2D)}{4M}$$

ς_s = shear in psi = 48,000 psi
d = diameter of axle bolt = 0.5 inches
D = travel distance during impact = 0.09 inches
M = motor vehicle weight = 3,500 lbs. F

Then:

$$v_1^2 = \frac{(48,000 \text{ lb f/in}^2)(\pi)(0.5 \text{ in})^2 (2 \cdot 0.096 \text{ in})}{4(108.7 \text{ lbs M})}$$

Where:

$$\frac{M}{g} = \frac{3,500 \text{ lbs}}{32.2 \text{ ft/sec}^2} = 108.7 \text{lbs} \cdot \text{sec}^2/\text{ft} = \text{slug}$$

$v_1^2 = 16.6 \text{ lb F} \cdot \text{in}$

$$v_1^2 = (16.6 \text{ lb F} \cdot \text{in})\left(\frac{32.2 \text{ ft}}{\text{sec}^2}\right)\left(\frac{12 \text{ in}}{\text{ft}}\right) = 6,414.2 \text{ in}^2/\text{sec}^2$$

$v_1 = 80$ in/sec = 6.7 ft/sec = 4.5 mph

This is the conceptual estimated peak speed at which the motor vehicle struck the cycle axle bolt, thus causing the shear.

The above shear equations were developed to give the forensic engineer a starting point for estimating collision speeds where shear is apparent on the cycle. It should be emphasized that these equations are not absolute. As more work is done in the laboratory, it is expected that the equations will be revised. Again, the shear equations were only intended to be a starting point in an accident investigation.

Conclusion

The equations in this chapter have been field tested and are accurate to a reasonable conceptional degree. They can be used at the initial stages of an accident investigation. Field verification should be conducted whenever possible. In addition, the unique parameters of the investigation should be quantified in the field.

Chapter 14

A Determination of Braking Distance for Cyclists In Emergency Stopping Situations

Introduction

During a bicycle accident reconstruction it is often necessary to ascertain the distance that a cyclist has available to bring the bicycle to an emergency stop. It is, therefore, the purpose of this analysis to provide the means by which such distance may be ascertained. The analysis will center on the caliper brakes that are typically used on most street and mountain bikes.

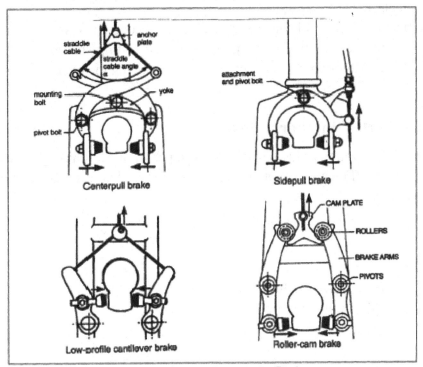

Figure 1 *Rim brake types. (Illustration by Rob Van der Plas)*

Deceleration Limits on Braking Distance

During the derivation of braking distance formulas, it must be emphasized that a cycle's front brake provides over 90 percent of the total retarding force at a deceleration of 0.5g. Also, a deceleration of 0.5g is the maximum that a cyclist can risk on level ground, in a crouched position before he/she goes over the handle-bars. Very little deceleration is needed to rotate a cyclist around the handlebars by raising the cyclist's center of gravity vertically upwards to a critical height. A one to two inch increase in the center of gravity is all that is necessary to rotate the cyclist over the handlebars. This means that a cyclist can only slow down at a rate of 4.91 m/sec^2 or 11 mph/set. A motor vehicle does not have the deceleration limits of a cycle. It can theoretically brake to the limit of tire-to-road friction. Thus, if a sudden braking situation arises on a road surface with, for example, a coefficient of friction (COF) of 0.8g, the motor vehicle can stop at a rate of 17.5 mph/set, while the cycle can only stop at a rate of 11 mph/set. The cycle has far less stopping capability (37.5 percent) than the motor vehicle.

The 0.5g value for cycle decelerations is a good conceptual value. To determine the actual deceleration value in a laboratory, testing should be done on the stopping capability of the subject cycle rims utilizing the cyclist's center of mass.

Coefficient Braking Formulas

While the deceleration values give a good upper limit of stopping distances, the pertinent coefficients of friction can also be used. This formula assumes no rear wheel lift:

$$D = \frac{V^2}{30(C_A + C_R)}$$

Where:
  V = initial velocity (mph)
  C_A = coefficient of adhesion
  C_R = rolling resistance
  D = distance in feet
  C = wheel circumference = nd
  d = wheel diameter

Typical values are;

	C_A	C_R
Dry Asphalt	0.85	0.014
Wet Asphalt	0.55	0.014
Ice	0.15	0.014

Example 1: If a cyclist enters an intersection of dry asphalt at a velocity of 20 mph, what is his stopping distance such that stability of the cycle is maintained?

$$D = \frac{(20 \text{ mph})^2}{30(0.85 + 0.014)} = 15 \text{ feet}$$

therefore:

$$D = \frac{v^2}{25.9} \text{ @ 20mph}.$$

Using the maximum deceleration rate such that stability of the cycle is maintained:

$$D = 50C + \frac{v^2}{30(C_A + C_R)}$$

Where:
  C = wheel circumference
  Note that a deceleration factor of 11 mph/set must be maintained. herefore, stopping time would be 1.82 seconds.

Coefficient Braking Formula on Wet Surfaces

Wet conditions obviously increase a cycle's stopping distance. This issue has been studied by the bicycle industry, as well as independent researchers. The studies indicate that the wheels of a cycle stopping on a wet surface at 10 mph will turn approximately 50 times prior to stopping. During these rotations, the COF between the brake blocks and rim returns to pre-wet conditions. This recovery rate has been found in field tests to be true for speeds between 10 - 20 mph. The Forensic Engineer should test recovery rates for speeds over 20 mph. A conservative formula for reconstructing stopping distances under wet conditions can be derived as follows:

$$C = \pi d$$

then:
   50C = distance traveled prior to recovery

$$D = 50C + \frac{V^2}{30(C_A + C_R)}$$

Where:
   V = initial velocity (mph)
   C_A = coefficient of adhesion on wet asphalt = 0.55
   C_R = rolling resistance
   C = wheel circumference in feet
   d = wheel diameter, feet

Example 2: If a cyclist riding a cycle with wheel diameter of 27 inches applies brakes at 20 mph on wet asphalt, the distance traveled would be:

$$D = (50)\left(\pi \frac{\times 27}{12}\right) + \frac{(20)^2}{30(0.55 + 0.014)} = 377 \text{ feet}.$$

This is the maximum stopping distance when the rim-brake combination experiences negligible COF until complete recovery.

If we assume a wet COF, which is one tenth of the dry COF, then an additional conceptual formula for stopping under wet conditions can be derived. This formula simply substitutes the wet coefficient of adhesion into the braking formulas as follows:

$$D = \frac{V^2}{30(C_A + C_R)} = \frac{V^2}{30(0.55 + 0.014)} = \frac{V^2}{16.9}$$

This formula more accurately represents braking conditions on wet asphalt.

Stopping With the Rear Brake Only

Many bicycle accident reconstructions deal with rear braking only. This is due to the improper use of both sets of caliper brakes, as well as the fact that some cycles, such as ungeared recreational bicycles, do not come equipped with front caliper brakes. Derivation of the formula for rear end braking is as follows:

Given that the COF between the tire and the road is 0.8

   R_r = perpendicular reaction force to the rear wheel

Then the maximum retarding force = 0.8 x R,

Assume the wheel base = 1.067 m

The moment about the front wheel becomes at M = 0.8 as;

Rr x 1067 mm + MRr x 1,143mn-r = 890 N x (1,067 - 432)mm

Where:

1,143 mm = perpendicular distance from the center of gravity of the cyclist to the ground.

Solving for R, = 285.2 N (64.1 lb F)

Using the classical force equation;

$$F = \frac{m}{g}a$$

$$a = \frac{Fg_c}{M} = \frac{-\mu R_r g_c}{N}$$

$$\frac{a}{g} = \frac{-0.8 \times 285.2N}{890N}$$

$$\frac{a}{g} = -0.256$$

$$a = -0.256g$$

The negative value denotes deceleration.

Note that the minimum deceleration to project a cyclist over the front of the cycle was found to be -0.56g.

Given that:

V_1 = -aT and at 20 mph as -29.4 ft/sec,

$$T = \frac{V_1}{-a} = \frac{-29.4 \text{ft/sec}}{-(0.256)(32.2 \text{ft/sec}^2)} = 3.56 \text{ sec}$$

$$S = \frac{V_1 + V_c}{2}T$$

Assume a complete stop. Then;

$$S = \frac{V_1}{2}T = \frac{29.4 \text{ft/sec}}{2} \times 3.56 \text{sec} = 52.3$$

A Summary of the Braking Distance Issue

The above formulas and relevant engineering literature reveal several conclusions about cycle braking distances in emergency stopping situations. These conclusions are as follows:

1. A deceleration above -0.5 g or -11 mph/set will likely result in a cyclist being thrown over the front of the cycle. Deceleration is not, however, a factor in rear braking only situations. It is virtually impossible for a cyclist to be thrown over the front of the cycle from application of the rear brake only.

2. The distance a cycle travels after the front and rear caliper brakes are applied is a function of the COF, converted to the coefficient of adhesion. Simply stated, the stopping distance of wet rims is almost twice that of dry rims. If the rims do not recover prior to 50 revolutions, the distance can be 25 times greater. Further verification in the laboratory is recommended if rim recovery is a factor in the accident reconstruction.

3. Many researchers contend that braking distances should only be determined at the subject accident site. However, the above equations have been verified in the field and found to be very accurate.

Acknowledgements

A substantial amount of data compiled from field investigations and by cycling companies was examined in connection with this analysis. Since much of the data is proprietary information, it is not cited here. I am, however, able to acknowledge my considerable reliance upon the excellent work of Whitt and Wilson[1]. These men are regarded as the fathers of bicycle engineering. They have managed to take engineering dynamics and apply the field to bicycle movement.

Endnotes

1. Whitt, Frank Rowland and Wilson, David Gordon, Bicycling Science, MIT, 1982.

Chapter 15

A Determination of the Structural Integrity of Bicycle Frames Subjected to Frontal Static Force

Introduction

During the course of bicycle accident reconstructions, subject cycle frames are far too frequently examined by untrained individuals who draw erroneous conclusions from their observations. For example, these individuals often conclude that top and down tubes bent back towards the frame were the cause of the accident under investigation (See Figure 1). As a result of this conclusion, the bicycle manufacturer is usually brought into a civil action based on the allegation that the bicycle was defective. After observing such events for a number of years, I decided to test a representative number of bicycle frames by subjecting them to a frontal static impact. Static testing was performed to achieve accurate results for comparison.

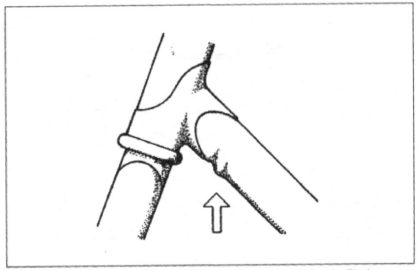

Figure 1 Checking for downtube damage. (Illustration by Rob Van der Plas)

Engineering Analysis

In order to obtain a representative sample of cycle frames, a wide range of frames were obtained from American and Taiwanese manufacturers. The only criteria utilized in the selection of the frames was that they be designed for use as all terrain bicycles (ATB) and be of a "general use" nature. This criteria was chosen because bent top and down tubes are usually found on the type of cycle frame specified. The reason for this coincidence has not yet been determined. However, it can be inferred that the type of rider that buys these frames tends to be inexperienced and more likely to engage in behavior that leads to front impact accidents.

Application of Frontal Static Force

As noted in Figures 1 and 2, a jig was constructed capable of applying a static orce to the front forks of the individual frames. A gauge calibrated in pounds per square inch (PSI) was used to record the force applied to the forks. In addition, the tests were videotaped to record the force applied, define the actual failure made (from directly above the jig), and provide a close up of the top and down tubes when failure was realized. The video tapes were spliced together after each test run so all views of the testing process could be viewed simultaneously.

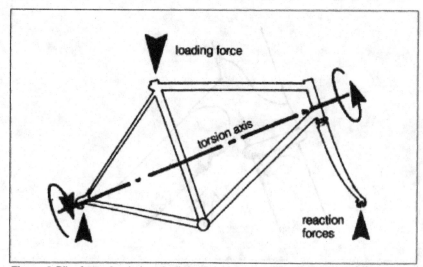

Figure 2 Bike frame loaded statically and dynamically. (Illustration by Rob Van der Plas)

For the purpose of this testing, failure of the cycle frame was defined as the stress point at which the metal exhibited plastic behavior. The static force was gradually applied until the gauge ceased to climb. This represented the maximum static force that could be applied immediately prior to failure, or the frame metal becoming plastic. As the frame collapsed, the gauge showed a sudden return to zero.

The Sequence of the Frame Collapse During the Application of Static Force

As the force was applied to the front fork, the sequence of the frame collapse was as follows:
1. Force was applied linearly and recorded on a force gauge.
2. Application of force was noted in PSI until the moment prior to collapse.
3. The force gauge ceased to climb and the gauge pressure was recorded.
4. As the frame collapsed, or became plastic, the force gauge reverted to zero.

Results of Frontal Frame Collapse Testing

The results of the frame failure test are noted in Figure 3. The failure pressure, defined in PSI, is converted from gauge pressure to failure force in pounds. The cycle models were listed on the chart in the order of testing. There is no particular significance to the order. A comparison of the test results (by use of a bar chart) is illustrated in Figure 4. The tested frames are noted in Figure 5 in their post-failure position.

Conclusions From the Frontal Static Testing

The results of this testing permit some interesting conclusions to be drawn about the design capabilities of the tested ATB bicycle frames. These conclusions are as follows:
1. The static force required to achieve a collapse ranged from 196 pounds to 439 pounds. This static force should not be confused with dynamic force which allows for deceleration of the frame upon impact. Decelerated force can be as much as 500 times greater. The static force testing conducted here is intended for comparative purposes only.

2. Analysis of the static force required to collapse a bicycle frame reveals that the same magnitude of force is needed between different ATB frames. Also, a frontal impact at a relatively low speed can cause a frame collapse. Dynamic testing bears this conclusion out and will be detailed elsewhere. This conclusion does not in any way mean that the design of ATB bicycle frames is defective. It does, however, mean that no bicycle frame can withstand a decelerated frontal impact. Examples of this type of frontal impact are a cyclist running into the side of a motor vehicle or losing control and riding into a drainage culvert.
3. Analysis of the static force test results clearly shows that the bend typically seen on the top and down tubes is the result of a frontal impact and, thus, is not an accident causal factor. Unfortunately, many self-proclaimed reconstructionists who lack any scientific or technical training will continue to opine that cycle companies are liable for accidents involving bent top and down tubes. Such an opinion is simply not based on fact. All cycle frames will bend when subjected to a frontal impact of sufficient magnitude.

Acknowledgements

The frames of Taiwanese manufacturers were obtained directly from the factory by Mr. Ken Justice. The jig noted in Figure 1 was constructed by Dr. Anand Kasbekar. The testing was also carried out in cooperation with Dr. Kasbekar. All work efforts were accomplished under the direction of James M. Green, P.E., DEE.

Figure 1 Overview of Frame Jig Used for Static Frame Tests

Figure 2 Testing and Recording Equipment Used in the Front Load Static Test

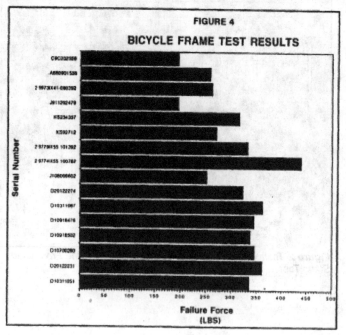

FIGURE 3

BICYCLE FRAME TEST RESULTS

SERIAL NUMBER	MODEL	FAILURE PRESSURE (psi)	FAILURE FORCE (lbs)
D10311051	DODSUN CATO	428	336
D20122231	DODSUN	460	361
D10700280	DODSUN	440	346
D10918502	DODSUN COYOTE THREE 4130 CROMO	430	338
D10918476	DODSUN COYOTE THREE 4130 CROMO	439	345
D10311087	DODSUN	461	362
D20122274	DODSUN	412	324
J108006602	JAZZ LATITUDE	321	252
2-9774X55-100782	TIMBER MOUNTAIN	559	439
2-9779X55-10392	TIMBER MOUNTAIN	423	332
K592712	CHALLENGER	345	271
K5234337	MERIDIAN RALLYE	403	317
J911202479	PARAGON CLASSIC	250	196
2-9973X41-060392	MURRAY STREET CYCLE	335	263
A660901536	AT 1200 80806	330	259
C9C332556	AT60	250	196

FIGURE 4

BICYCLE FRAME TEST RESULTS

Figure 5 Individual Collapsed Frames

Figure 5 *Individual Collapsed Frames*

Figure 5 Individual Collapsed Frames

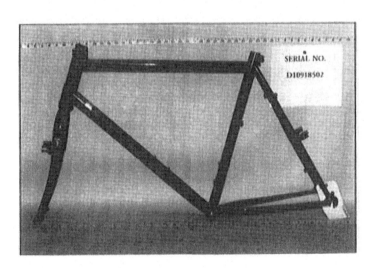

Figure 5 Individual Collapsed Frames

Figure 5 Individual Collapsed Frames

Figure 5 Individual Collapsed Frames

A Determination of the Structural Integrity of Bicycle Frames

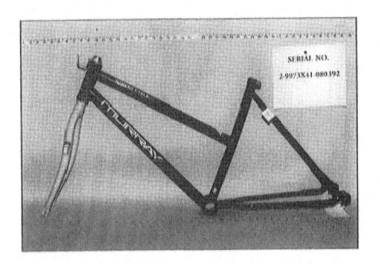

Figure 5 Individual Collapsed Frames

Chapter 16

The Engineering Dynamics of a Cyclist Being Thrown over the Front of a Cycle During a Sudden Stop

Introduction

In reconstructing bicycle accidents, untrained, so called "experts" often observe bends in the top and down tubes and come to some amazing conclusions. One such conclusion is that the cycle was defective and the defect caused the rider to fall. This opinion, which cannot be substantiated by the engineering dynamics, leads to unnecessary litigation against bicycle companies. The purpose of this analysis is to derive the equations that describe a cyclist being thrown over the front of a cycle during a sudden stop. In addition, this analysis will illustrate the sequence of events that result from this type of fall, including the point at which the top and down tubes actually bend. The equations described in this analysis have all been verified in the laboratory and by peer review from fellow professional engineers.

Engineering Analysis

Figure 1 shows a conceptual drawing of a cycle with the items needed for the analysis. Pertinent points are as follows:

- C = The center of gravity of the cycle rider system. This can be visualized as a point centered between the cyclist's pelvic girdle at approximately the navel. This analysis focuses on the direction the C travels under various conditions.
- H = The height the C must travel in order for the cyclist to continue over the front of the cycle in a forward fall. In some of the engineering literature, H is expressed as ΔH, or the difference in C when the cyclist is in the seated position and when the cyclist meets the critical height as a prelude to a forward fall. ΔH becomes a factor once rotation starts and is illustrated in position 3.

Figure 1 *Example of a rider beginning forward rotation. (Photo from The Mountain Bike Book by Rob Van der Plas)*

The derived equations will use various approaches to C_g and H to define the minimum speed necessary for a cyclist to go over the front of the cycle. The equations utilize various locations on the cycle rider system. These locations are found on Figure 1. The actual dynamics of rotation are shown in positions 1 through 5. Figure 1 and positions 1 through 5 are located at the end of this paper. An examination of Figure 1 reveals that an arc is drawn around This represents a cyclist rotating around the handlebars when enough to allow the C_g to rotate around the handlebars.

Determining ΔH can only be done with a computer program such as "ADAM" or the equivalent.[1] The physiological measurements of the cyclist and the exact measurements of the subject cycle should be used with the program. Most ΔH's on level ground are approximately one to two inches for a typical mountain or road cycle.[2] The drawings of the cyclist rotating around the handlebars (positions 1 through 5) illustrate very clearly how the ΔH and C_g move when a cycle is brought to a sudden stop.

Testing on the various chrome-moly steel cycle frames on the market revealed that an average force of 250 pounds static force is necessary to bend the top and down tubes. This force value will be used for deriving the appropriate equations.

The Kinetic Energy Method

Using the concept that, upon sudden stopping of the cycle, all energy is kinetic, the following equation can be used:

$$1/2 MV^2 = MgH$$

This can be simplified to;

$$1/2 V^2 = gH$$

where:
- M = mass of rider
- g = gravitational constant
- H = ΔH or critical height of C_g
- V = velocity of cyclist

Simplifying;

$$V^2 = 2g\Delta H$$

$$v = [2g\Delta H]^{0.5}$$

Assume that ΔH has been found to be 1.75 inches, the critical velocity where the cyclist will go over the front of the cycle is;

$$V = \left[2 \times 32.2 \text{ ft/sec}^2 \times \frac{1.75}{12} \text{ feet}\right]^{0.5}, \text{where } V = 3.06 \text{ ft/sec} = 2.1 \text{ mph}$$

Fork Deflection Method

If it is assumed that the front fork is deflected during the impact, thus absorbing energy, then the energy balance becomes:

$$1/2 M\Delta V^2 = MgH + FD$$

where:
- F = static force needed to deflect the fork to the point of plasticity.
- D = Distance the fork moves prior to deformation or plasticity.

To obtain F and D, laboratory testing should be conducted to accurately measure deformation. In the field it is difficult to obtain laboratory verification of fork deflection, so representative values may have to be used. If the mass of the cyclist is 190 pounds, then

$$\frac{1}{2} [190 \text{ lbs}] \Delta V^2 = [\ 190 \text{ lbs} \times 32.2 \text{ ft/sec}^2 \ \frac{1.75}{12.0} \text{ ft}\ ] + \left[\frac{250 \text{ lbs}}{2} \times \frac{4.375}{12} \text{ ft}\right]$$

Since force applied must go from zero to 250 pounds, it should be averaged by dividing by two. The AV involved in the expression;

$$\Delta V = V_1 - V_F$$

where:

V_1 = initial speed just prior to point of impact.
V_F = final speed which is assumed to be zero.

Therefore, the equation becomes;

$95V^2 = 892 + 45.57$
$V^2 = 9.87$
$V = 3.14 \text{ ft/sec} = 2.13 \text{ mph}$

As expected, slightly more speed is necessary in the fork deflection method than in the kinetic energy method to have the cyclist rotate around the handlebars.

The Straight Arm Method

Using conservation of momentum, a derivation can be found where the cyclist rotates around the handlebars. (Note that conservation of momentum equations are not discussed here. The derivation should be self-explanatory. If, however, additional information is desired, it is recommended that the reader consult a standard engineering dynamics text.)

Therefore;

If $H_i = H_F$
$H_i = MV_iL_1 + I_AW_1$ (See figure 1)
and $H_F = I_AW_F$
@ $\Delta H = 0$

$$H_i = H_F \text{ and } V_iL_1 = I_AW_F$$

$$W_F = \frac{MV_iL_1}{I_A}$$

If $\Delta E = O$
Then;

$$\frac{1}{2} I_A W_F^2 = MgH$$

Substituting;

$$\frac{1}{2} I_A \left[\frac{MV_i L_1}{I_A}\right]^2 = MgH$$

$$I_A \left[\frac{MV_i L_1}{I_A}\right]^2 = 2MgH$$

Solving for V_i;

$$V_1^2 = \frac{2MgH I_{AI}}{M^2 L_1^2}$$

And

$$V_1^2 = \frac{2I_A gH}{ML_1^2}$$

$$V_1 = \left[\frac{2I_A gH}{ML_1^2}\right]^{0.5}$$

Since $I_A = I + Mr^2$ and $I = 0$

Then

$$V_1^2 = \left[\frac{4Mr^2 gH}{ML_1^2}\right]^{0.5} = \left[\frac{4r^2 gH}{L_1^2}\right]^{0.5}$$

$$V_1 = \frac{2r}{L_1}[gH]^{0.5}$$

where:
  r = distance from the cyclist's center of gravity (C_g) to the front fork tips
  L_1 = distance from the cyclist's center of gravity (C_g) and the bottom bracket center

Given that;
  r = 2.5 feet
  L_1 = 2.91 feet

$\Delta H = 4.375$ feet'

The minimum velocity that can throw the cyclist over the front of the cycle during a sudden stop is;

$$V_1 = \frac{2r}{L_1}[g\Delta H]^{0.5} = \frac{2 \times 2.5\,\text{ft}}{2.91\,\text{ft}}\left[32.2\,\text{ft/sec}^2 \times \frac{4.375}{12}\,\text{ft}\right]^{0.5}$$

$$V_1 = 5.89 \text{ ft/sec} = 4.0 \text{ mph}$$

Method Comparison

Comparing the velocity required to rotate a cyclist over the front of a bicycle yields the following data:

Method	Velocity (ft/sec)	Velocity (mph)
Kinetic Energy	3.06	2.00
Fork Deflection	3.14	2.13
Straight Arm	5.89	4.00

Kinematics of the Actual Fall

Using the three methods developed, the actual mechanics or kinematics of a fall over the front of a cycle can be illustrated using a computer simulation. The kinematics reveal that the major driving force in the forward fall is the direction the center of gravity (C_g) goes when the necessary ΔH is reached. This ΔH is reached when the bicycle stops abruptly (i.e., as a result of striking an object or a premature release of the bike's front wheel which causes the front fork crown to drop onto the wheel). When the latter occurs, the front fork tips will often fall behind the wheel axle. This is equivalent to stopping the cycle suddenly and transferring a tremendous amount of force to the frame.[3] Among the many computer simulations that are available to simulate these types of falls, there are some apparent constants. These constants are:

1. Once the cyclist's center of gravity is raised above the critical ΔH, the cyclist will rotate around the handlebars in some fashion. The

usual mode of rotation is with the arms fairly rigid and the hands affixed to the handlebars throughout the sequence.
2. The cycle will follow behind the body of the cyclist, since the larger mass of the cyclist will dictate the direction and sequence of the fall.
3. The critical height of the center of gravity (ΔH) is a function of the topography that is being ridden. To accurately determine the ΔH, a computer simulation should be done with the exact dimensions of the cyclist, the cycle and the surveyed terrain. General values of ΔH can be utilized in conceptual accident reconstructions as follows:

Rider is descending a hill, $\Delta H = 4\text{-}5$ inches
Rider is on level ground, $\Delta H = 1\text{-}2$ inches

Again, an illustration of the sequence of these types of falls is noted in positions 1 through 5.

Conclusion

Damage to a bicycle's top and down tubes is the result of a sudden stop, not the causal factor of a sudden stop. The force vectors on a cycle during normal riding are exactly opposite of the vectors needed to bend the top tube, down tube and front fork. It is thus apparent that any "expert" who opines that the cycle is defective because the cyclist was thrown over the front from a sudden stop is unaware of the physics involved in the accident.

Endnotes

1. For this study, the subject bicycle was modeled Super 3D from Silicon Beach Software. using the three-dimensional CADpackage,
2. Green, James M., P.E., DEE, *A Determination of the Structural Integrity of Bicycle Frames Subjected to Frontal Static Force,* submitted to the Taiwan International Bicycle Exposition,
Taipei, Taiwan, April, 1993. [This analysis was based on a fairly extensive study in which a testing apparatus was constructed and over thirty frames were deformed. For purposes of reconstructing bicycle accidents, the values that can be used are; D = 4.375 inches and F = 250 pounds static force. See Chapter 15 for a summary of this test procedure.]
3. Force of impact onto a bicycle frame during these types of falls is described in another study. That study shows that tremendous amount of decelerated force is transmitted to the cycle's top and down tubes as the cycle pivots around the front drop outs. See, Green, James M., P.E., *Determination of the Velocity of a Cyclist from Deformation of the Cycle from a Frontal Impact,* submitted to the Taiwan International Bicycle Exposition, Taipei, Taiwan, April, 1993. See Chapter 23 for a summary of the results.

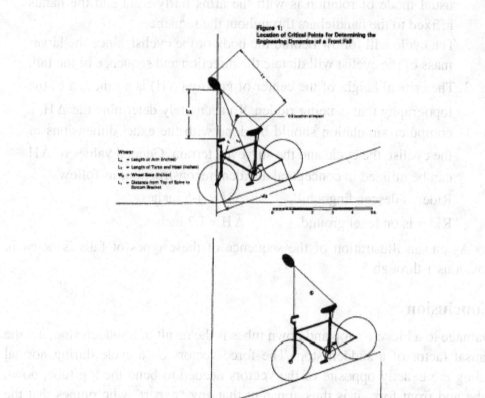

Figure 1:
Location of Critical Points for Determining the Engineering Dynamics of a Front Fall

Where:
L_a = Length of Arm (inches)
L_t = Length of Torso and Head (inches)
W_b = Wheel Base (inches)
L_s = Distance from Top of Spine to Bottom Bracket

Position 1

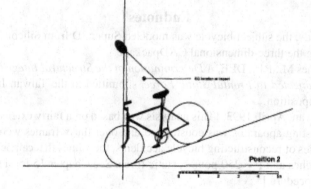

Position 2

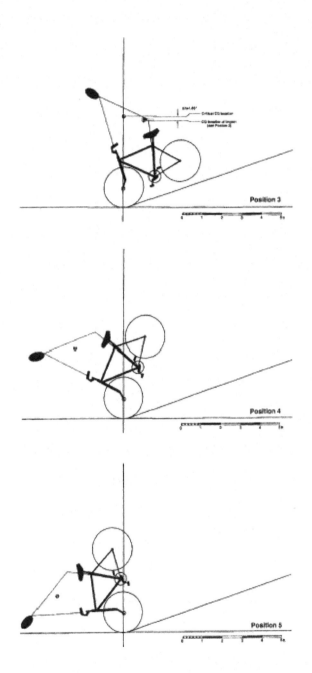

Chapter 17

A Method for Determining Impact Damage to a Helmet When the Cyclist is Thrown Over the Front of the Cycle During a Sudden Stop

by Anand David Kasbekar, Ph.D.[1] and James M. Green, P.E., DEE

Introduction

The equations developed in the previous chapter describe the kinematics of a cyclist being thrown over the front of a bicycle during a sudden stop. This type of fall occurs when, for any number of reasons, there is sufficient deceleration to cause the cyclist's center of gravity to rotate past the vertical plane through the bicycle's front axle, thus causing the cyclist to rotate over the handlebars. For example, such falls may occur when the cycle's front wheel prematurely releases, Chapter 7, or the cyclist rides over a poorly designed or constructed railroad crossing, Chapter 8. During reconstructions of such falls, the damage sustained by a bike helmet is often critical for determining the cycle speed prior to impact. This chapter describes a test apparatus which has been developed for collecting bicycle helmet crush data and to study the effects of impact velocity with regard to helmet damage.

Test Apparatus

The pendulum type test apparatus shown in Figure 1 simulates rotational velocity which a rider experiences during a forward fall from a cycle. A side view of the apparatus is shown in Figure 2. Weights are positioned on the pendulum arm to simulate the cyclist's center of mass during the test. A locking universal joint has been incorporated to simulate neck articulation for the purpose of accurately orienting the head form. The manner in which the head form is attached to the apparatus is shown in Figure 5. The length of the pendulum arm can be adjusted to achieve the appropriate impact velocity of the head form. When the pendulum is released, the head form strikes a moving surface of asphalt or concrete with a coefficient of friction that matches the condition at the accident site. The interaction between the helmet and the ground is simulated using a hydraulic ram that moves the surface at a predetermined speed. This movement is shown in Figure 3. The ram used to move the pad is shown in Figure 4.

In configuring the test apparatus, it was assumed that 100 percent of translational motion is converted to rotational motion during the fall. Note, however, that, in most falls including those caused by the premature release of the cycle's front wheel, some degree of translational motion exists and this translational component is dissipated during the forward sliding of the cycle and rider on the pavement. A method for accurately simulating this dissipation has not yet been developed. However, field observations reveal that for low speed accidents the translational velocity at impact is nominal and probably has no significant effect on the impact force experienced by the bike helmet.

Determining the Approximate Impact Velocity for a Cyclist Pivoting About the Front Wheel

To achieve the correct impact velocity with the test apparatus, the pendulum arm must be adjusted so the head form and rider center of mass are at the appropriate heights. This critical height may be obtained from the following equations. These equations relate impact velocity to the initial height of the rider's center of mass.
Given:

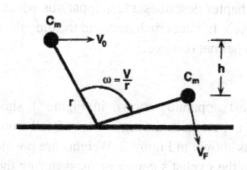

where:
    V_0 = initial velocity
    V_F = impact velocity
    h = change in the height of the rider's center of mass
    E_0 = initial energy of system
    E_F = final energy of system
    C_m = rider center of mass
    ω = angular velocity

Assume no skidding.

Method 1

$E_0 = E_F$ (conservation of energy)

$E_0 = 1/2 m V_0^2 + mgh$

$1/2 m V_0^2 + mgh = 1/2 m V_F^2$

$V_0^2 + 2gh = V_F^2$

$V_F = (V_0^2 + 2gh)^{1/2}$

Method 2

$E_0 = E_F$

$E_0 = 1/2 m V_0^2 + mgh$

$E_F = 1/2 I \omega_F^2$

$1/2 m V_0^2 + mgh = 1/2 I \omega_F^2$

Assume point mass @ $C_m \Rightarrow I = mr^2$

$1/2 m V_0^2 + mgh = 1/2 (mr^2) \omega_F^2$

$V_0^2 + gh = r^2 \omega_F^2$

$\omega_F^2 = \left(\dfrac{V_0^2}{r^2} + \dfrac{gh}{r^2} \right)$

Since $\omega = \dfrac{V_F}{r}$

Therefore,

$\left(\dfrac{V_F}{r} \right)^2 = \dfrac{V_0^2}{r^2} + \dfrac{2gh}{r^2}$

$V_F = (V_0^2 + 2gh)^{1/2}$

Note that the result is the same equation as derived with Method 1.

Assume, for example, that a cyclist traveling at 10 mph is thrown over the front of the cycle due to a sudden stop. The change in C_g height is equal to 1.2 feet, as measured from basic percentile dimensions of adults in the United States.[2]

The speed of impact of the rider's center of mass will be:

$V_F = (V_0^2 + 2gh)^{1/2}$

$V_F = [(14.7 \text{ ft/sec})^2 + 2(32.2 \text{ ft/sec}^2)(1.2 \text{ ft})]^{1/2}$

$V_F = 17.1 \text{ ft/sec} = 11.7 \text{ mph}$

$\omega_{C_g} = \dfrac{11.7}{r_{C_g}}$

Where:

r_{C_g} = distance from pendulum pivot to C_g

Assuming that the body behaves as a rigid mass, then:

$\omega_{head} = \omega_{C_g}$

$\dfrac{V_{head}}{r_{head}} = \omega_{C_g}$

$V_{head} = \omega_{C_g}$

$V_{head} = \dfrac{11.7}{r_{C_g}} r_{head}$

where r_{head} = distance from pendulum pivot to center point on head

Conclusion

The pendulum test apparatus described in this chapter may be used to simulate with relative accuracy the damage sustained by a bike helmet during a forward fall from a cycle. In conjunction with equations provided, the cyclist's impact velocity can be approximated in most cases. Lastly, the apparatus may be used to test and compare helmets subjected to various impact conditions.

The development of this test apparatus and the collection of helmet data is an ongoing process with the intent of improving flexibility, ease of use and accuracy of this device. The investigator may obtain various final impact speeds by adjusting the height of the apparatus head form.

Endnotes

1. Dr. Kasbekar is a consulting engineer for Research Engineers, Inc. and President of Visual Sciences, Inc., Raleigh, NC.
2. Woodson, Wesley E., *Human Factors Design Handbook,* Second Edition. McGraw Hill, 1992, pages 541-542.

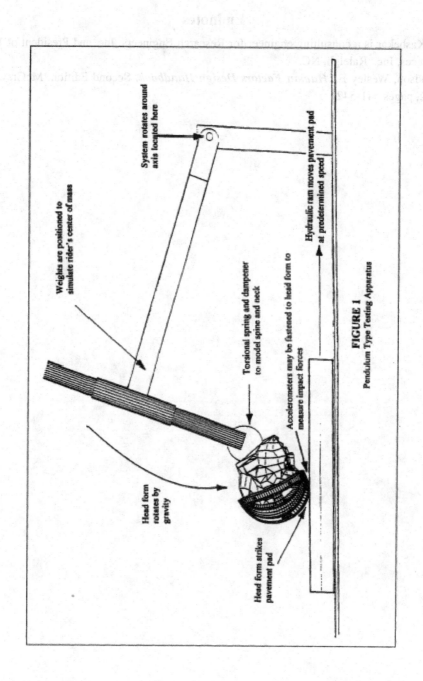

FIGURE 1
Pendulum Type Testing Apparatus

Figure 2 Testing Apparatus After Release of Head Form

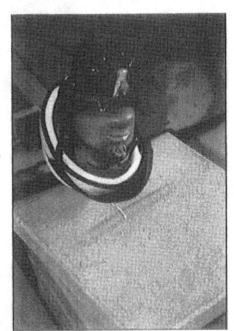

Figure 3 Head Form and Test Helmet as Pad is Moved Under Helmet

Figure 4 Hydraulic Ram in Lower Right Corner Pushes Pad to Simulate Sliding

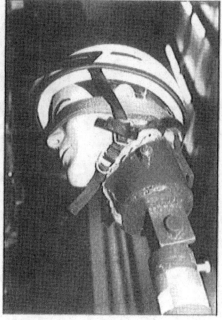

Figure 5 Head Form Secured to Adjustable Neck Joint

Chapter 18

A Determination of the Correct Method for Developing Motor Vehicle Speed of Impact in Bicycle Accident Reconstruction Collisions Using Engineering Literature and Field Verification

Introduction

There has been a great deal of confusion among forensic engineers regarding the relationship between linear carry or throw distance and motor vehicle impact with cyclists. The primary reason for the confusion is that the cyclist is an elastic body and the motor vehicle is a rigid body. The engineering literature prior to the 1970's tended to treat the two entities as rigid bodies. As a result, substantial error was introduced into the calculations of vehicle speed. The purpose of this chapter is to summarize the developments in the engineering literature concerning the calculation of motor vehicle speed from the known linear throw distance of the cyclist. The results of the analysis may be applied to pedestrians as well. But note that the literature cited herein is careful to distinguish between pedestrian and bicyclist throw distances. Further, throw distances are usually determined in the field by using human cadavers on a bicycle or separately to simulate pedestrian impacts. The work described in this chapter will, however, refer to both cyclists and pedestrians, unless specifically stated otherwise. After review of the literature, a recommended method for determining motor vehicle speed will be defined.

Engineering Literature

It should be noted that the definitive work in this area first addresses impacts with pedestrians. This is due to the fact that the engineering community has only recently focused on cyclist throw distances. Such developments will be examined chronologically.

The Schmidt Method

Schmidt[1] did the first definitive work on the development of motor vehicle speed of impact from victim throw distance. Using pedestrians as a focal point and one actual intersection collision, Schmidt emphasized characteristics of the impact on the victim, as well as vehicle damage and dimensions.

According to Schmidt, acceleration in the Y direction is given as:

Equation 1

$$Y = Y_0 + V_{Y0}[T - T_0] - 1/2g[T - T_0]^2$$

where:
- Y = Height of center of mass (G_m) or vertical component
- V_{Y0} = Vertical velocity component at time, T_0
- T_0 = Time at initial point of impact (POI)
- T = Time at any point on the trajectory

Assuming no air resistance and that the X-component of velocity is constant up to time, T, the distance traveled becomes:

Equation 2

$$d_1 = V_{X0}[T_1 - T_0]$$

where:
- V_{X0} = horizontal velocity component
- d_1 = horizontal distance traveled at time, T, while air borne

Schmidt then adds a skidding velocity component and combines horizontal and vertical distances vectorially to yield the initial speed component;

Equation 3

$$V_{X0}(h) = [f^2(h) + e]^{0.5} - f(h)$$

where:

$$f(h) = \mu\sqrt{2g}[\sqrt{h} + \sqrt{h + Y_0}]$$

(Equation 4)

- μ = coefficient of friction (COF)
- $e = 2\mu g d_t$
- Y_0 = value of Y at time t

The various components of the trajectory are noted in Figure 1. Converting the pedestrian center of mass to the cyclist center of mass renders these equations suitable for a cyclist trajectory.

Example Calculation Using the Schmidt Method

Given the following parameters on a sideways collision:

$\mu = 0.7$

$g = 32.2$ ft/sec^2

$Y_0 = 3.0$ ft (measured center of mass of the cyclist above the plane, P)

$d_r = 133.5$ ft (linear distance cyclist is thrown)

Therefore:

$$\mu \sqrt{2g} = 0.7 [64.4 \text{ ft/sec}^2]^{0.5} = 5.6 \text{ ft}^{0.5}/\text{sec}$$

$$e = 2mgd_T = (2)(0.7)(32.2)(133.5 \text{ ft})$$

$$= 6018 \text{ ft}^2/\text{sec}^2$$

$$f(h) = 5.6 \text{ ft}^{0.5}/\text{sec} [\sqrt{h} + \sqrt{h+3\text{ft}}]$$

If h = 0 (see figure 1), then;

$$f(h) - 5.6 \text{ ft}^{0.5}/\text{sec} \sqrt{3 \text{ ft}} = 9.70 \text{ ft/sec}$$

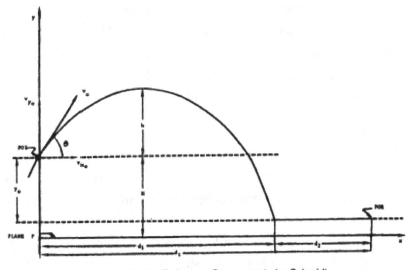

Figure 1 Pedestrian and Cyclist Trajectory Components by Schmidt

Substituting to determine the horizontal velocity using Equation 3 yields:

$$V_{X0}(0) = [(9.7 \text{ ft/sec})^2 + 5950 \text{ ft}^2/\text{sec}^2]^{0.5} = 77.7 \text{ ft/sec}$$

$$V_{X0}(0) = [77.7 \text{ ft/sec}][1.47 \text{ ft/sec/mph}]^{-1} = 52.89 \text{ mph}$$

If h = 10 ft (see figure 1) then;

$$f(10) = 5.6 \text{ ft}^{0.5}/\text{sec} \, [\, \sqrt{10 \text{ft}} + \sqrt{13 \text{ ft}} \,] = 37.8 \text{ ft/sec}$$

Then;

$$77.7 \text{ ft/sec} - 37.8 \text{ ft/sec} = 39.9 \text{ ft/sec}$$

Substituting:

$$V_{X0}(0) = (39.9 \text{ ft/sec})(1.47 \text{ ft/sec/mph})^{-1} = 27.1 \text{ mph}$$

for the speed of the motor vehicle.

The values for various ranges of horizontal velocities are noted in Table 1.

Critique of the Schmidt Method

The Schmidt method was an excellent first attempt at quantifying linear carry distance of a pedestrian as a function of motor vehicle speed at the point of impact. The pertinent areas of interest are as follows:

a. The fact that Y_0 is defined as the center of mass or gravity (G_m) at the POI enables the forensic engineer to easily convert from the G_m of a pedestrian to the G_m of a cyclist. To convert the center of gravity (G_m), multiply the wheel base by 1.071. Take this value and place the G_m on a point on the plumb line from the ground through the bottom bracket. For example, in Figure 2, the wheelbase is 42 inches yielding a center of gravity of 45 inches on a plumb line through the bottom bracket (42 inches x 1.071 = 45 inches).

b. The primary problem with the Schmidt method is that the engineer must iterate the solution using various values of h_o. This makes the eventual solution procedure cumbersome and yields a probable range of motor vehicle speeds rather than an exact speed.

The Schmidt method is a very sound procedure for determining motor vehicle speed for pedestrian and bicycle reconstruction and should be utilized within the limits just described.

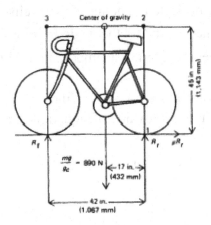

Figure 2 Determination of Center of Mass of the Cycle Rider System

The Searle Method

The Searle Method[2] employs an equation for the projection velocity associated with a given total trajectory. The equation enables the engineer to bracket the limits within which the projection velocity must lie. Projection velocity is the speed of impact of the motor vehicle at the point of impact. The Searle Method also uses hard data on frictional coefficients in the final formula.

The formula for a total trajectory, including bouncing and sliding is:

Equation 5

$$V = \frac{\sqrt{2\mu g s}}{(\cos\theta + \mu \sin\theta)}$$

where:

- g = gravitational constant (32.2 ft/sec^2)
- μ = coefficient of friction
- θ = projection angle in degrees
- s = flight distance where $s = \frac{2}{g}[uv]$
- u = horizontal component of the flight distance or $v = \cos\theta$

It is interesting that the Searle method does not contain the coefficient of rebound since it was to be taken into account under the original premise of the equation. This omission is likely due to the fact that the rebound coefficient is very small. Also, the possible bouncing of the cyclist is negligible in comparison to the total distance, S, unless the angle θ is very low. In addition, coefficient of friction values, μ, are not as important as skid marks for determining motor vehicle speed.

If the angle of projection is not known, then upper and lower limits must be determined. For the minimum value of Equation 5, a new derivation from the calculus shows that:

$$\text{Tan } \theta = \mu \text{ so that } \theta = \arctan \mu$$

Equation 6

$$V_{min} = \left[\frac{2\mu gs}{1 + \mu^2}\right]^{0.5}$$

For the upper limit of velocity the derivation yields:

Equation 7

$$V_{max} = [2\mu gs]^{0.5}$$

This assumes that the angle of projection is below some critical value θ. Projection equations will derive θ to be 45°. However, when the coefficient of friction becomes a factor, then:

$$\theta_{crit} = 180° - 2 \arctan \frac{1}{\theta}$$

The coefficient of friction, μ, can be plotted against the critical projection angle, θ_{crit}, using these relationships. This plot, as noted in Figure 3, enables the forensic engineer to determine upper and lower velocity limits in an accident where skidding or bouncing of the cyclist is a factor.

Example Calculation Using the Searle Method

If the coefficient of friction at an accident site is 0.40 on dry asphalt, the critical projection angle then becomes 44°. If the linear throw distance is found to be 40 feet, the velocity ranges are calculated as follows:

$$V_{min} = \sqrt{\frac{2\mu gs}{1 + \mu^2}} = \sqrt{\frac{2 \cdot 0.4 \cdot 32.2 \cdot 40}{1 + (0.4)^2}}$$

$$V_{min} = (29.8 \text{ ft/sec})(1.47 \text{ ft/sec/mph})^{-1} = 20.3 \text{ mph}$$

$$V_{min} = \sqrt{2\upsilon gs} = \sqrt{2 \cdot 0.4 \cdot 32.2 \cdot 40}$$

$$V_{max} = (32.1 \text{ ft/sec})(1.47 \text{ ft/sec/mph})^{-1} = 21.8 \text{ mph}$$

Searle goes on to define projection efficiency as a function of different types of vehicles. An additional series of graphs are provided utilizing field data generated by Appel.[3]

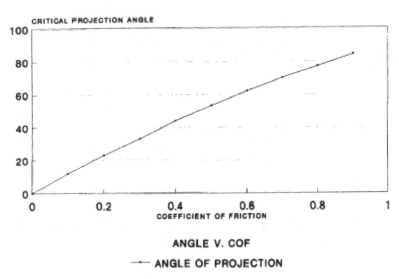

Critique of the Searle Method

The Searle method is extremely sound mathematically. The method also takes into account the bounce or slide of the cyclist or pedestrian as a function of coefficient of friction. The pertinent points about the Searle method are as follows:

1. The Searle method is the only one that accurately determines the bouncing and sliding part of a trajectory. This distinction is not a trivial one since the "skipping" part of the trajectory can be greater than the airborne section.
2. In applying the Searle method to bicycle accident reconstruction, it should be noted that a cyclist's center of gravity is higher than that of a pedestrian. The higher center of gravity will result in a longer

linear distance in the trajectory. A review of the throw distances developed by Appel reveals that the difference in cyclist and pedestrian center of gravity is treated as insignificant. This is due to the addition of the "skipping" distance in the derivation. The added distance more than compensates for any nominal difference in the center of gravity distance.

Of all the methods studied, the Searle method is the most mathematically-sound and correlates well with field data.

Endnotes

1. Schmidt, David H. and Nagal, Donald A., *Pedestrian Impact Case Study,* Proceedings of Fifteenth Conference of the American Association of Automotive Medicine, October 20-23, 1971, pages 151-167.

2. Searle, John A. and Searle, Angela, *The Trajectories of Pedestrians, Motorcycles, Motorcyclists, etc., Following a Road Accident,* (1983) SAE Paper No. 831622.

3. Appel et. al. provided graphs of actual impact accident speeds for different types of motor vehicles and throw distances at the 2nd International IRCOBI Conference in 1976. These graphs are a part of the Searle paper and correlate well with the maximum and minimum impact speeds derived above.

TABLE 1: CYCLIST, PEDESTRIAN-TRAJECTORY PARAMETER RANGE USING THE SCHMIDT METHOD

h Maximum Height Above Ground (ft)	h Peak Height of Parabola (ft)	T_i Time Airborne (sec)	μ Coefficient of Friction	$X_o + d$ Distance (ft)	a Distance Sliding (ft)	V_x Horizontal Component of Velocity (ft/sec)	V_x Horizontal Component of Velocity (mph)
3	0	0.43	0.7	29.5	104.0	68.2	46.5
			0.4	13.0	110.5	53.2	36.3
			0.2	16.7	116.8	38.7	26.4
5	2	0.91	0.7	54.5	79.0	59.6	40.6
			0.4	43.6	89.9	47.7	32.5
			0.2	32.7	100.8	35.8	24.4
7	4	1.16	0.7	64.5	69.0	55.6	37.9
			0.4	52.6	80.9	45.4	30.9
			0.2	40.2	93.3	34.6	23.6
9	6	1.36	0.7	71.7	61.8	52.7	35.9
			0.4	59.5	74.0	43.6	29.8
			0.2	45.7	87.8	33.6	22.9
11	8	1.54	0.7	77.2	56.3	50.2	34.2
			0.4	65.1	60.4	42.3	28.9
			0.2	50.4	83.1	32.7	22.3
13	10	1.70	0.7	82.0	51.5	48.1	32.8
			0.4	69.0	64.5	40.6	27.7
			0.2	53.2	53.2	31.9	21.8

Notes: $Y_o = 3$ ft
$d_f = 133.5$ ft

Chapter 19

Field Testing to Determine Throw Distance - The Haight and Eubanks Studies'

Introduction

Linear throw distance has been studied in the field by Rusty Haight and Jerry Eubanks. In the Haight and Eubanks field studies, various motor vehicles were used to strike an exemplar bicycle with a dummy cyclist at various speeds. An example of the field testing is shown in Figures 1 through 6. The results of this analysis is shown in Table 1.

Haight and Eubanks define the throw distance as the distance from the point of impact to the final uncontrolled point of rest for the cyclist. The entire distance travelled, including any portion spent on the striking vehicle, or in the air, is included in the throw distance noted in Table 1. This definition of throw distance includes the slide distance which is the distance the body travels on the ground as it comes to rest. The throw distance of both the bike and the body is a straight line to the center of mass of the object from the point of impact.

The bicycle heading angle was recorded in Table 1 as the angle of the bicycle in relationship to the vehicle's heading.

Conclusion

The author's comparison of the Haight and Eubanks studies reveals that the throw distance correlates well with the Searle method, but not the Schmidt method. Future work is needed to develop a good empirical formula from field data to better define throw distance. Until such work is accomplished, it is recommended that the the field can be field verified by Haight and Eubanks. Searle method be used and verification, where possible, be done using data in this chapter. It is also possible that the actual accident conditions can be field verified by Haight and Eubanks.

Figure 1 Motor Vehicle Approaches Cycle (Photo provided by Peter Murphy of the Massachusetts State Police)

Figure 2 Initial Impact With Cyclist Being Vaulted Onto the Hood (Photo provided by Peter Murphy)

Field Testing to Determine Throw Distance

Figure 3 As the Motor Vehicle Stops, the Cyclist is Ejected From the Hood and Windshield (Photo provided by Peter Murphy)

Figure 4 The Cyclist Continues the Ejection Trajectory Onto the Ground (Photo provided by Peter Murphy)

Figure 5 Final Resting Place of the Cyclist and Cycle (Photo provided by Peter Murphy)

Figure 6 Crush Data and Windshield Damage to Subject Motor Vehicle (Photo provided by Peter Murphy)

Table 1

A Determination of Throw Distance of a Cycle and Cyclist at Various Impact Speeds

Test	Location	Vehicle	Speed (mph)	Throw Distance Bicycle	Throw Distance Rider	Bike Hdg. (degrees)
1	San Diego	1984 Toyota Corolla	21	82.5	53	0
2	San Diego	1984 Toyota Corolla	18.4	35.5	24	0
3	San Diego	1984 Toyota Corolla	19.6	35	28.5	0
4	San Diego	1984 Toyota Corolla	27.5	52.5	82.5	0
5	San Diego	1984 Toyota Corolla	20.7	44.25	51	0
6	San Diego	1984 Toyota Corolla	16.9	38.25	30	0
7	San Diego	1984 Toyota Corolla	18.6	43.75	36	0
8	San Diego	1984 Toyota Corolla	17.6	48	43.5	0
9	San Diego	1984 Toyota Corolla	21.3	52.5	51	0
10	College Station	1984 Mercury Montego	57	153	105	0
11	Ottawa	1986 Chevrolet Caprice	29.2	34[11]	40.75	90
12	Match AFB	1983 Datsun Pickup	38.5	89	67	0
13	Rosenberg	1986 Ford LTD	27.2	45.5	46	90
14	Rosenberg	1986 Ford LTD	32.5	104	59	300
15	College Station	1976 Ford F-150 Pickup	33.6	102	75.75	300
16	College Station	1987 Hyundai Excel	29.7	87	75.83	350
17	Kelowna	1983 Chevrolet Cavalier	31.3	43	47	0
18	Kelowna	1982 Dodge Pickup	29.5	46	41.3	70
19	Ocean City	1983 Mercury Capri	42.5	209	110	0
20	Plano	1985 Chevrolet Camaro	26.5	68	43.5	0
21	Plano	1979 Chevrolet Monte Carlo	30.4	75	69.5	0
22	College Station	1972 Chevrolet Concord Wgn	32.5	49.5[8]	67	0
23	College Station	1972 Chevrolet Concord Wgn	46.25	104[9]	54	270
24	Anchorage	1986 Ford LTD	42	73.75	69.6	270
25	Anchorage	1986 Ford LTD	37.9	72	73	0
26	Austin	1977 Pontiac Grand Prix	41.5	134	108	0
27	Austin	1979 Plymouth Horizon	42.1	61.75[10]	28.6	270
28	San Antonio	1983 Buick Skylark	42.6	135	73.6	270
29	San Antonio	1976 Chevrolet C-20 Pickup	30	50	49.8	270
30	Louisville	1979 Honda Accord	26.8	57.7	58	60
31	Louisville	1982 Honda Civic	35.0	70.6	60.5	180
32	Conroe	1991 Chevrolet Caprice	31.7	50.4	32.4	0
33	Conroe	1991 Chevrolet Caprice	34.7	67	64.1[1]	0

Test	Location	Vehicle	Speed (mph)	Throw Distance Bicycle	Throw Distance Rider	Bike Hdg. (degrees)
34	Conroe	1991 Chevrolet Caprice	34.7	78.9	67.6[2]	0
35	Austin	1976 Ford LTD	24.8	34.4	35.6[3]	45
36	Austin	1976 Ford LTD	24.8	34.4	33.5[4]	45
37	Austin	1976 Ford LTD	32.2	83.2	51.8	180
38	Seattle	1979 Ford Mustang	31.7	53.3	49.5	180
39	Seattle	1984 Chevrolet Caprice	37.3	93	72	0
40	New Braintree	1984 Renault Alliance	31.9	48	52	0
41	New Braintree	1982 Chevrolet Impala	34.0	76	64.7	270
42	Bergen County	1979 Ford Mustang	30.7	50.8	62.3	90
43	Bergen County	1979 Dodge Magnum	32.9	48	54	0
44	Agoura	1981 Honda Accord	40.5	90.3	78.6[5]	90
45	Agoura	1986 Chevrolet Van	30.7	49.8	54.8	0
46	San Diego	1988 Dodge Dakota	25.0	6	4[6]	0
47	San Diego	1988 Dodge Dakota	26.0	63	51	0
48	San Diego	1988 Dodge Dakota	19.75	41.5[7]	53	0
49	Charleston	1981 AMC Concord	42.5	72	104	45

(Data provided by Rusty Haight and Jerry Eubanks)

* Bicycle direction is indicated relative to the vehicle's direction of travel in degrees such that the longitudinal axis for the front of the car and the car's movement heading is "0 degrees." With this convention, the front of the bike facing passenger side of the vehicle is 90 degrees, the front of bike facing the striking vehicle such that the bike's front wheel is struck first is a relative 180 degrees.

[1] Small bicycle, simultaneous impact with large bicycle (note 2)

[2] Large bicycle, simultaneous impact with small bicycle (note 1)
The Conroe tests indicated as numbers 33 and 34 are actually one test involving two bikes

[3] Rider on seat

[4] Rider on handlebars
The Austin tests indicated as numbers 35 and 36 are actually one test involving one bicycle with one dummy on the bicycle's seat and one on the bicycle's handlebars

[5] Dummy caught on fifth wheel (throw distance effected by dummy hang-up on test vehicle equipment)

[6] Bicycle sideswiped by pickup, partial contact trajectory category

[7] Bicycle entrapped under pickup post collision

(8) Bicycle entrapped under car post collision
(9) Partial contact and bicycle entrapped under car post collision
(10) Partial contact
(11) Bicycle entrapped under car post collision

Endnotes

1. W.R. "Rusty" Haight and Jerry J. Eubanks, ACTAR certified accident reconstructionists and course instructors with the Texas Engineering Extension Service, Texas A&M University, have together conducted more than 115 crash tests involving pedestrian or bicycle crash dummy struck by a motor vehicle. The data included in this chapter reflects some of the documentation of 49 of those crash tests involving the dummy astride a bicycle struck by a motor vehicle

Chapter 20

A Determination of the Actual Point of Impact in Bicycle Accident Reconstruction Involving Motor Vehicles

Introduction

In a bicycle accident reconstruction, the actual point of impact (POI) plays a vital role in identifying the cyclist's position just prior to impact. In turn, the pre-impact position of the cyclist and other participants in the accident scenario can help define the causal factors of the accident. For example, if it is found that the cyclist suddenly veered into the zone of containment just before the POI with a motor vehicle, then the cause of the accident may have been the poor riding habits of the cyclist. The purpose of this analysis is to define some common axioms for finding the POI in motor vehicle/bicycle collisions. The analysis assumes that the cycle stays on the road surface and is not projected via a trajectory following impact. A trajectory analysis, which is examined in another study, contains engineering dynamics that are entirely different from those used for the determination of surface impacts. The dynamics of defining surface POI's will be explained using an actual field example.

Analysis of Surface Points of Impact

An overriding concern in the determination of surface points of impact (POIs) is the likelihood that the bicycle wheel will collapse upon impact. If the wheel collapses, the POIs may be altered.

A bicycle wheel is very durable when not impacted by an outside force. In fact, the bicycle wheel design allows substantial amounts of radial force to be transmitted to the wheel. Unfortunately, the design also allows complete and total collapse of the wheel when lateral force is applied (i.e., as when a cycle is hit by a motor vehicle).

If wheel collapse occurs upon impact, the cycle will usually contact the road at the crank outer change ring. On very rare occasions, the pedal arm will contact the road surface. Given the configuration of the pedal chain arm, the pedal arm will usually rotate until the chain ring contacts the surface. It is usually safe to assume chain ring contact unless inspection of the cycle reveals otherwise.

The wheel collapse and chain ring contact with the road surface may, as noted above, change the POIs. To understand how the change in POIs occurs, one must fast consider the various engineering physics involved.

A Case Study Example of POIs Determination

A cyclist was allegedly hit from the rear by a truck travelling at 45 mph. The cyclist claimed that he was riding on the fog line at the time of the accident. The truck driver said that the cyclist veered into his path of travel without warning, thus making the impact unavoidable. The collision point on the truck was the collapsed right front fender. Inspection of the cycle revealed that the rear wheel was collapsed. In addition, the left rear down tube chain stay was destroyed, indicating a trajectory blow of approximately 30'.

The police officer who investigated the scene located the POIs at a point almost directly on the fog line. This finding appeared to confirm the cyclist's contention that he was riding on the fog line. However, a secondary investigation of the damage to the cycle revealed that the cyclist's contention was incorrect.

The bicycle inspection revealed that the outer chain ring was collapsed and the left rear chain stay was broken. Also, the rear wheel was collapsed and the right front fork was bent to the left. This damage pattern indicated that the bike was struck on the left rear chain stay at an angle of 30° from the normal forward plane of travel. The investigating officer documented this fact, but neglected to consider in his analysis the time it would take the bike to collapse onto the large chain ring.

Determination of the Actual Point of Impact

The analysis considers the following:
- The velocity of the truck = 45 mph
- The distance of the chain ring to the road before impact = 7 inches

Therefore, the velocity of the truck (V_T) can be shown as:

$$V_T = (45 \text{ mph})(1.47 \text{ ft/sec/mph}) = 66.15 \text{ ft/sec}$$

The distance the chain ring will travel before striking the ground (D_{cr}) can be shown as:

$$D_{cr} = (7 \text{ inches})(12 \text{ inches/foot})^{-1} = 0.58 \text{ feet}$$

A Determination of Actual Point of Impact

The rate at which an object falls, under the influence of gravity, is given by the classic physical equation:

Equation 1;

$$S = V_0T + 1/2 AT^2 \text{ feet}$$

where:

V_0 = initial velocity (in this case, 0 ft/sec)
A = acceleration of gravity or 32.2 ft/sec^2
S = distance of fall in feet

To find the time, Equation 1 must be solved for the quadratic variable, T;

$$T = \frac{-V_0 + [V_0^2 + 2AS]^{0.5}}{A}$$

since $V_0 = 0$, Equation 2 can be solved to yield;

Equation 3;

$$T = \frac{-V_0 + [-V_0 + (2)(32.2 \text{ ft/sec}^2)(0.58 \text{ ft})]^{0.5}}{32.2 \text{ ft/sec}^2}$$

Solving Equation 3 yields;

$$T = \frac{[37.35 \text{ ft}^2/\text{sec}^2]^{0.5}}{32.2 \text{ ft/sec}^2} = \frac{6.11 \text{ ft/sec}}{32.2 \text{ ft/sec}^2} = 0.189 \text{ sec}$$

Therefore, 0.189 seconds is the time it would take the outer chain ring to strike the pavement.

Combination of the Velocity of the Truck and the Time of Crank Chain Ring's Fall

While the chain ring is falling, the truck continues to move at a velocity of 45 mph, or 66.15 ft/sec. The distance that the truck moves, while the chain ring on the cycle is falling, can be defined as follows:

Equation 4;

$$D_{45} = (66.15 \text{ ft/sec})(0.189 \text{ set}) = 12.5 \text{ feet}$$

Since the truck hit the cycle on the left rear chain stay, there are two points of impact at issue. The first POI is on the left rear chain stay. The second POI is on the chain ring. It is important to note that the chain stay, not the chain ring was hit by the truck. Yet, the chain ring POI is important since the chain ring will eventually strike the ground and leave the marks noted in the police report.

Determination of the Actual Point of Impact Versus the Point of Impact From the Accident Report

If one assumes a 30° right triangle, the point of impact (POI) can be defined in the following spatial relationship at a motor vehicle speed of 45 mph.

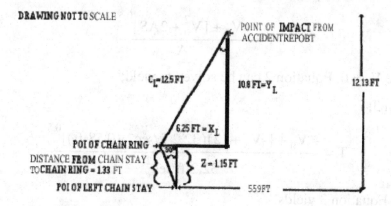

Figure 1 Relationship Between the Accident Report POI and the Actual POI

The equations for solving the actual points of impact are classic trigonometry.

Given that the distance of 12.5 feet was derived from Equation 4, then the right triangular relationship is:

Equation 5;

$$\sin \theta = \frac{X_L}{C_L} = \frac{X_L}{12.5}$$

which becomes;

$$\sin 30_o = \frac{X_L}{12.5}$$

and $X_L = 6.25$ feet

The distance back from the accident report POI is calculated as;

Equation 6;
$$\cos\theta = \frac{Y_L}{12.5}$$

$$Y_L = (12.5)(\cos 30°) = 10.8 \text{ feet}$$

The hypotenuse of the smaller triangle is the distance from the large chain ring to the chain stay. This value is measured from the cycle and is 1.33 ft. Again, trigonometry is used:

Equation 7;
$$Z = (\cos\theta)(1.33 \text{ ft}) = 1.15 \text{ feet}$$

and

$$X = (\sin\theta)(1.33 \text{ ft}) = 0.665 \text{ feet}$$

Conclusion

To determine where a motor vehicle strikes a cycle, several laws of physics must be considered:

1. Marks in the roadway will occur after the cycle has impacted the road some distance away from the actual POI. In rear end collisions, the rear wheel of the cycle is destroyed and the large chain ring is traumatized as it cuts into the roadway. To expect marks on a roadway to occur at the exact point of impact ignores the fact that the cycle will fall at acceleration rate of gravity.
2. Other types of collisions may occur when a motor vehicle strikes a cyclist. The cyclist can be hit from the side or from the front. In all instances, an inspection of the cycle can reveal the point at which the car hit the cycle and the point where the cyclist hit the pavement.

The equations contained in this chapter can be used to determine the time it will take a cycle to drop to the pavement and the distance in the X and Y direction that is appropriate for locating the actual point of impact.

Chapter 21

Determination of the Velocity of A Cyclist From Deformation of the Cycle From A Frontal Impact

Introduction

When a cyclist impacts an object, such as a motor vehicle, deformation of the cycle usually occurs. Deformation is often found approximately two inches from the head set on the top and down tubes. On other occasions, the fork crown may be damaged or the fork blades may be bent towards the down tube. In all instances, the fork tip ends are displaced back towards the down tube and bottom bracket. From this deformation it is possible to determine the cycle's speed at impact.

There are two types of force associated with a frontal impact. It is important that the forensic engineer knows the difference in these forces, because their misapplication will give incorrect impact speeds. Thus, derivation of the two types of forces is necessary to define the difference and to determine speed of impact.

Average Force - The force applied to the front fork during the total bending moment is the total average force of the deformation. This force measurement assumes that deformation occurs and that the frame completely absorbs the force prior to the cyclist being thrown over the front of the cycle.

Instantaneous Maximum Force - The force applied to the front fork during the total bending moment will have an instantaneous maximum point. This point is always several orders of magnitude higher than the average force. A plot of the dynamic force application over time will reveal the point. Instantaneous maximum force must be measured in the laboratory.

Derivation of Average Force Formula

Assume a cycle hits an object and the final velocity equals zero.

$$F = M \cdot A$$

$$F = M \frac{\Delta V}{\Delta T}$$

$$F = M\frac{V_1 - V_F}{\Delta T}$$

where:

V_1 = Velocity at impact

V_F = Final velocity ($V_F = 0$)

$$\Delta T = \frac{\Delta L}{\overline{V}}$$

where:

ΔL = Distance fork travels, or maximum deflection of front fork

$$\overline{V} = \text{Average Velocity} = \frac{V_1 + V_F}{2}$$

Then:

$$\Delta T = \frac{2\Delta L}{V_1 + V_F}$$

Substituting:

$$F = \frac{M(V_1 + V_F)}{\frac{2\Delta L}{V_1 + V_F}} \qquad F = \frac{M(V_1^2 + V_F^2)}{2\Delta L}$$

Assume that front fork deflection is measured in the laboratory by applying an average static force of 3,500 pounds to the front fork of an exemplar cycle. The test yields a front fork deflection of 0.75 feet prior to collapse of the top and down tubes. To determine the impact speed of a 210 lb. cyclist using the average force method:

$$F = \frac{M}{g}\frac{(V_1^2 + V_F^2)}{2\Delta L} \text{ where } V_F = 0$$

$$F = \frac{210 \text{ lbs}}{32.2\frac{\text{ft}}{\text{sec}^2}} \cdot \frac{V_1^2}{2 \cdot 0.75}$$

$$3500 \text{ lbs} = \frac{(6.52)V_1^2}{1.50}$$

$$V_1^2 = 805.2 \text{ ft}^2/\text{sec}^2$$

$V_1 = 28.3$ ft/sec $= 19.3$ mph or the conceptual estimated impact speed of the cyclist hitting a perpendicular object.

Numerous frontal impact tests in the laboratory reveal that an average force of 3,500 lbs. is the norm. The tests further reveal that fork deflection varies with velocity in the following relationship;

$$3500 = \left[\frac{V_1^2}{2\Delta L}\right] 6.50$$

or

$$539 = \left[\frac{V_1^2}{2\Delta L}\right]$$

simplifying to;

$$1078 = \frac{V_1^2}{\Delta L}$$

or

$$V_1 = \frac{[1078 \Delta L]^{0.5}}{1.47} = \text{impact speed (mph)}$$

Note that it is recommended that an exemplar bike frame be tested in the laboratory to determine actual force of impact and cyclist impact speed. However, this is not always possible and a conceptual impact velocity must sometimes be obtained during the course of the accident reconstruction.

The average force method is an excellent way for the forensic engineer to estimate speeds of impact. By measuring the fork deflection linearly, accurate velocities at impact can be obtained. Laboratory data indicates that this method yields a deviation from actual impact speed of only three percent. The average force method is, thus, very accurate and may be used with a high degree of confidence when circumstances preclude laboratory testing.

Derivation of the Instantaneous Maximum Force Formula

An exemplar fork must be tested in the laboratory to determine the maximum force applied during dynamic loading on the front fork. During the testing, the fork is driven into a pad that can record the amount of force applied over the deceleration period. In addition, the cycle is allowed to "bounce" back from the testing pad so that the force is not absorbed by the frame. This represents a cyclist hitting an object perpendicular to his path of travel before rotating over the front of the cycle.[1] The instantaneous maximum force can be extrapolated from the test printout. The instantaneous maximum force is only one small segment of decelerated force. The value of this force will be higher than the average force values described above. The derivation of the instantaneous maximum force equations can be accomplished using the classic force of impact equations:

$$F = M \cdot A$$

$$F = \frac{M}{g} \cdot \frac{\Delta V}{\Delta T}$$

$$\Delta T = \frac{\Delta L}{\overline{V}}$$

$$\overline{V} = \text{Average Velocity} = \frac{V_1 + V_F}{2}$$

where:
- m = Mass of cyclist
- A = Gravitational acceleration
- ΔT = Time of instantaneous deceleration
- ΔL = Fork deflection
- $\overline{V}$ = Average velocity

Substituting into the force equation;

$$\Delta T = \frac{2\Delta L}{V_1 + V_F}$$

and substituting $\overline{V}$ yields;

$$F = \frac{M}{g}\left[\frac{V_1^2 + V_F^2}{2\Delta L}\right]$$

An example calculation assumes the following:
- Impact speed = 51.5 ft/sec
- Residual speed = 7.5 ft/sec
- Fork deflection = 0.75 feet

Rider mass = 210 pounds

The calculation becomes;

$$F = \frac{M}{g}\left[\frac{V_1^2 + V_F^2}{2 \cdot 0.75}\right] = \frac{210 \text{ lbs}}{32.2 \text{ ft/sec}^2}\left[\frac{(51.2)^2 + (7.5)^2 \text{ft/sec}}{2 \cdot 0.75 \text{ ft}}\right]$$

F = 11,642 pounds instantaneous decelerated force

This method has some draw backs in determining cycle speed at the point of impact. The instantaneous maximum force tends to be unique to the type of frame being ridden. As a result, the engineer should test the subject frame in a dynamic mode to define the force.

Conclusion

Test comparisons of the average force and instantaneous maximum force methods have shown that the former is better for predicting a cyclist's speed at impact. It must be emphasized that this analysis assumes a measurable, permanent deflection of the cycle front fork.

Endnotes

1. Green, James M., P.E., DEE, *The Engineering Dynamics of a Cyclist Being Thrown Over the Front of a Cycle During a Sudden* Stop. Submitted to the Taiwan International Bicycle Exposition, Taipei, Taiwan, 1993.

Chapter 22
A Determination of Bicycle Speed By Using Crush Data

Introduction

In connection with their efforts to improve vehicle safety, car manufacturers have studied for a number of years the relationship between the crush of a motor vehicle after impact and the velocity of the vehicle at the time of impact. As a result of these studies, crush data is now generally available for most motor vehicle models. This data can usually be used by the forensic engineer to determine the often critical issue of pre-collision motor vehicle speed. A review of the pertinent literature reveals that, with the exception of a study by Renault (attached), similar crush data is not currently available for bicycles. In an effort to expand upon the Renault study, an extensive study was conducted on the relationship between the crush of numerous exemplar bicycles after impact and the velocity of the cycles at the time of impact. The crush data study was conducted in connection with the previously discussed *Johnson* accident reconstruction (see Chapter 9). The facts of the subject accident are highlighted below.

The crush data study included the construction of a test track and the subsequent testing of over eighty (80) bicycles. The testing consisted of crashing loaded bicycles into a steel barrier at various velocities and measuring the distance of the front fork deformation. This distance was then correlated with velocity of impact. From this correlation, the speed of impact of the accident cycle was determined. The study revealed that the correlation between cycle velocity and front fork deflection distance was extremely high. As such, the method of crush data compilation utilized in the study and discussed in detail below can be used successfully to determine pre-collision cycle speed in other accident reconstructions.

Facts of the Accident[1]

A cyclist who was descending a five degree sloped hill at night was seriously injured when a 1988 Jeep Wrangler turned in front of the cyclist at an intersection. The cyclist struck the right rear of the Jeep with a force sufficient to drive him through and out the other side of the Jeep. The cyclist subsequently brought suit against the bicycle manufacturer. Other defendants had minimal or no insurance and, thus, were not a factor in the plaintiff's case. In his lawsuit, the cyclist claimed that the cause of the accident was the cycle manufacturer's failure to install a light on the bicycle.

Given the above facts, the accident reconstruction involved an analysis of the pertinent conspicuity issues. In addition, the speed of the cyclist was also examined to determine the potential existence of contributory negligence on the part of the cyclist.

Engineering Analysis

A test track was constructed using a series of 40 foot sections of W6 x 90 steel beams welded end to end and set onto cross ties. A drawing of the track is shown in Figure 1.

Test Track

An exact replica of the 1988 Jeep Wrangler was constructed at the end of the track and anchored in the ground. The replica was offset 15 degrees for Jeep movement laterally across the cycle's direction of travel.

A motorcycle rear wheel was converted into a spool onto which the tow rope attached to the cycle was wound. The power source employed was a 1981 Honda CB750K motorcycle. By using a tachometer and the appropriate gear ratio, the predicted maximum towed vehicle speed prior to impact could be determined. The actual speed of the cycle prior to impact was determined using a radar gun with an accuracy of plus or minus one mile per hour.

Bicycle

Several exemplar bikes were obtained from the manufacturer for testing. Since none of the bike components contributed to the deformation of the frame and fork, the only components that were included in the cycle set-up were the headset, handle bars, saddle and stem. This set-up enabled the bikes to be equipped with top and bottom weights for stability on the track.

For further stability, the bicycles were placed on a sled which was the object that was directly accelerated along the track. A pair of front axle extenders were manufactured to allow the yoke from the sled to hook over the extenders. The extenders also prevented the wheel from releasing upon impact.

The cycle was loaded in a manner that simulated the distribution of a 178 pound cyclist's weight 60 percent over the rear wheel and 40 percent over the front wheel. This was accomplished by attaching the unweighted cycle to scales with ropes looped around the front and rear wheels. Weights were then applied throughout the cycle to obtain the necessary weight distribution.

Dynamic Testing on a Track

Each cycle was fitted with the distributed weights and accelerated into the side of the exemplar Jeep. The vector definition of the angle of impact is as follows:

Assume that the Jeep speed is 8 mph and the top speed of the cycle is 40 mph on a 5 degree slope; then the lateral movement of the Jeep across the cycle's path can be shown as:[2]

Top speed of cycle = 40 mph · 1.47 mph/ft/sec = 58.8 ft/sec

Testified speed of Jeep = 8 mph · 1.47 mph/ft/sec = 11.8 ft/sec

Addition of the vectors can be shown as;

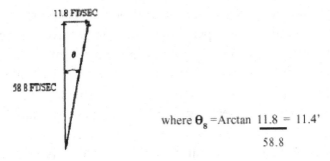

where $\theta_8 = \text{Arctan } \dfrac{11.8}{58.8} = 11.4°$

If the Jeep had a potential high end speed of 10 mph or 14.7 feet second, the addition of the vectors can be shown as;

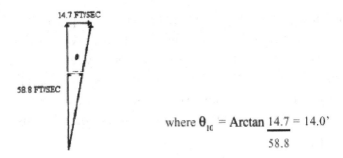

where $\theta_{10} = \text{Arctan } \dfrac{14.7}{58.8} = 14.0°$

A conservative angle of 14 degrees was then used between the cycle and the side of the exemplar Jeep in the following manner

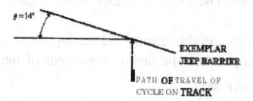

A distinction must be made between a perpendicular collision, as defined by the physical spatial relationship of the cycle and motor vehicle, and the speed vectors involved in the collision. In the former, the motor vehicle may cross the path of the cycle in a manner that is not, in fact, perpendicular. However, when determining impact speed, the direction of travel between the cycle and the motor vehicle (position vectors) is not important as compared to the speed of the two bodies. In other words, speed, not physical angle of impact, is important when determining the proper angle adjustment for the test barrier.

The above distinction arises from the fact that the cycle has a point of impact against the side of the motor vehicle of approximately one (1) inch (width of the cycle's tire). Given this minimal impact point, if the motor vehicle crosses the front of the cycle at some unknown angle, the more conservative approach is to adjust the barrier angle by utilizing speed vectors, rather than position vectors.

Obviously, it would be better to actually know the position vectors involved in the cycle/motor vehicle collision and then transpose position and speed into one vector. However, this information cannot always be obtained from the accident scene. Thus, the laboratory proven axiom for cycle/motor vehicle accidents is that speed or velocity vectors take precedence over position vectors. However, the forensic engineer should, where possible, use both the position and speed vectors.[3]

After the track was constructed, as noted in Figure 1, exemplar cycles were loaded with weights as previously described . The loaded cycles were accelerated into the barrier at the velocities noted on Fi.gure 2 . The deformation of the cycles' front forks was then measured. The basic measurement consisted of the difference from the plumb line that passes through the center of the axle prior to impact and the centerline of the axle after impact.

Analytical Results and Applicable Court Testimony

A plot of the fork deflection is shown in Figure 2. By conducting a linear regression, a curve was generated for extrapolating velocity of impact. The curve reveals a correlation coefficient with a 98 percent level of confidence. Given this level of confidence, the fork deflection on the subject cycle can be compared with the crush data on the exemplar cycles to obtain the subject cycle's impact speed. For the *Johnson* accident, a speed of 40 mph was extrapolated from the graph. This velocity was significant given the fact that the speed limit at the accident intersection was 35 mph.

Opposing counsel frequently seeks to have valid scientific data excluded from evidence by employing a line of questioning which suggests that the data is not based on a commonly accepted method of testing. Experience has shown that this is particularly true with respect to crush testing performed for the purpose of determining speed of impact. It is, however, hoped that this analysis will help defeat such challenges in the future by clearly showing that crush testing is, indeed, a commonly accepted method for determining speed of impact. To further this goal, attached as an addendum is the French Paper (Renault study) that utilizes crush data to determine speed of impact. In addition, the results of further crush testing are shown in Table 1 and plotted in Figure 3. These results are highly consistent with those reflected in Figure 2. It is therefore apparent that the data in Figure 2 is quite valid for estimating pre-impact speed.

Conclusions of Crush Data Testing and Determination of Pre-impact Cycle Speed

In reconstructing bicycle accidents, the pre-impact speed of the subject cycle helps to determine such critical issues as reaction times, line of sight, and conspicuity. Some salient points about the use of crush data to determine cycle pre-impact speed are as follows:

1. Since the use of a test track and crush data may not always be inancially practical, the algorithm in Figure 2 can be used to determine conceptual pre-impact speed with barriers or motor vehicles. When employing Figure 2 in this fashion, an algorithm should be generated from data on the accident cycle to verify the onceptual speed results. It is understood that, without access to a working test track, the generation of this algorithm may be difficult?
2. Conventional road and mountain bike frames with some type of front fork rake were used in the crush testing. Eleven additional

frames of different conventional design were tested and the fork deformationcompared to the algorithm in Figure 2. The speed and fork deformation relationships were within five percent of the values produced with the algorithm in Figure 2. These relationships were apparent with different types of tubing (i.e., tubing of varying thickness and materials). The algorithm in Figure 2 is an accurate method for estimating cycle speeds in perpendicular intersection collisions.
3. No testing was done on forks equipped with hydraulic damping systems, such as the Rock Sox or equivalent. As a result, the method of crush data development and analysis described in this paper only applies to conventional fork blades with classic blade rakes.

Conclusion

The automobile industry has used crush data for a number of years to determine speed of impact. The results achieved by the industry have as high degree of reliability. The crush testing discussed above indicates that the same level of confidence can be attained with bicycle crush data.

A Determination of Bicycle Speed by Using Crush Data

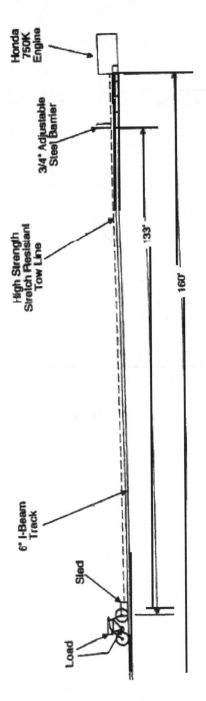

Figure 1 Bicycle Test Track

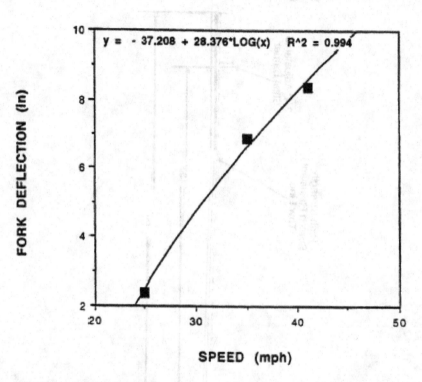

Figure 2 Speed vs. Fork Deflection

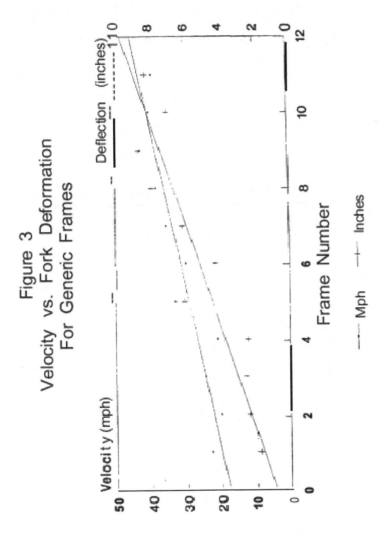

Figure 3
Velocity vs. Fork Deformation
For Generic Frames

Frames used were common 56 cm street frames of 4130 chromium - molybdenum steel.

Table 1

Speed vs. Fork Deflection Using Generic Frames

FRAME	MPH	DEFLECTION (inches)
1.0	23.0	1.8
2.0	20.0	2.4
3.0	24.5	2.55
4.0	21.0	2.44
5.0	33.0	6.1
6.0	30.0	4.3
7.0	35.5	6.2
8.0	38.1	7.9
9.0	42.5	8.6
10.0	40.0	7.0
11.0	39.0	8.2

ADDENDUM

INTERNATIONAL MEETING ON THE SIMULATION AND RECONSTRUCTION OF IMPACTS IN COLLISIONS (1978)
Lyon, France

METHODOLOGY OF RECONSTITUTION OF ACTUAL ACCIDENTS ON TEST TRACKS
D. Criton, P'. Ventre
Direction of Research and Development
RENAULT National Car Company

Definition of Test Parameters

There are only three parameters which may be used to reconstruct an accident: velocities, angles of impact and points of impact.

Parameters are defined by the methods developed for detailed analysis of accidents (1 and 2):

- Velocity variation (AV) by use of the energy dissipated during a collision by the vehicle or vehicles with the obstacle or obstacles.
- Impaction of vehicles and closeness of centers of gravity are determined by distortion of vehicles and obstacles after the impact.

Once the basic data are retained, test parameters are selected by strictly respecting approach velocities and angles for frontal impacts, for the important factor is to obtain the same velocity variations as during the actual accident. For all impacts in which two velocity vectors at an angle are compounded, the important factor is to obtain the correct resulting vector in order to ensure that the trajectory followed by the occupants of the impacted vehicle inside their compartment is faithful.

Practical Achievement of Test (3)

General Layout

Tests are carried out on the test and collision track of the Lardy Technical Centre, which is designed to meet all present and future standards but also to reproduce any type of accident.

This last requirement is the reason for a layout composed of two distinct zones (Figure 1).

The first zone (A) with an impact wall of 60 m^3 of reinforced concrete, i.e. 150 tons, is the traditional section where all tests are carried out against an infinitely rigid obstacle. This may be the wall itself with or without a dynamometric buffer, but is may also be an inclinable wall which may vary through the whole range of 0 to 45° (mainly 30°) to the left or to the right, a dynamometric pole or any other form of obstacle required to reconstitute an actual accident.

The second zone (B) is located at the intersection of the main 350 m track with a 100 m track. This latter cuts the first at right angles at a distance of 200 m from the impact wall and provides a number of roms of collisions: frontal, lateral, from behind, tests of roll-over, etc. A third oblique track allows for impacts at various angles over a quarter of a circle of radius 50 m from the point of crossing of main tracks. Each track, of width 7 m of particular section (two 1% gradients), converges towards the center of a hangar of area 1500 m^2 (50 m x 30 m) which provides freedom from the effects of bad weather. In this place, also, tests are carried out with pedestrians and cyclists.

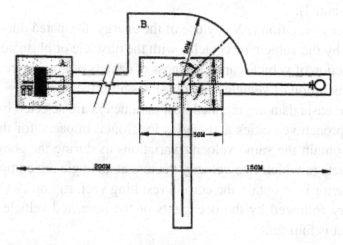

Figure 1 *Overall View of Collision Test Tracks*

Both zones are fitted with equipment to provide best utilization of tests carried out on them. A wide bridge is placed in front of the wall and also at the intersection to facilitate overhead photography and a wide pit is provided for low level photography.

The bridges also accommodate all the lighting systems so that there can be no impediment of any sort to the movement of the vehicles after impact. The two pits are fitted with projectors built into recesses to provide the widest possible field of vision for cameras filming the undersides of vehicles.

Furthermore, in order to improve the quality of photographs which are particularly interesting in relation to deformation of structures the pits are covered by thick laminated glass sheets resting on a wide mesh metal lattice (600 mm x 300 mm); this enables the pits to be completely closed and provides a surface on which vehicles may move almost normally.

The Winch

Vehicles are drawn at speed by a winch which comprises a set of two 300 CV IC engines which may be used separately or simultaneously. A thick cable (dia 18 mm), runs unterruptedly along the center of the main track. It transmits the drive force by means of a carriage which can be attached to it as required at any point of the track by means of a special gripper.

Accuracy of Trajectory

Though the vehicles are accelerated by two IC engines each fitted with an automatic gear box with three forward speeds and one reverse speed, velocity at the point of impact is regulated electrically by means of a Foucault current, Telma-type truck brake. Data on the vehicle velocity is supplied to the mini-computer by a tachymetric dynamo which is directly connected to the cable and varies the intensity of the current transmitted to the braking mechanism so that the braking mechanism cancels excess power of the engines to establish an equilibrium when the desired velocity is attained (Figure 2). The two IC engines are controlled by prior setting of their carburetors.

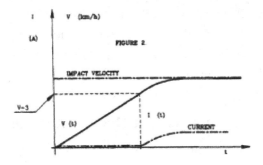

This method has two advantages:
- errors of impact velocity are less than 1%,
- force of traction on the vehicle is always positive because it is at least the load necessary to preserve the velocity acquired; this latter advantage is essential for good fidelity of vehicle trajectory.

Though a continual positive traction force is a necessary condition to provide accuracy of trajectory, the following precautions must nevertheless be taken:
- state of track must be impeccable: the 1% half gradients which converge towards the rail facilitate flow of rain water,
- the front wheel assembly of the vehicle must be completely set on a VISUALINER bench and tires must be checked,
- the connection between vehicle and carriage provided by a V cable the loop of which marks the center line of the vehicle,
- measurement cables are balanced on either side of the vehicle.

When these precautions are taken, vehicle trajectories are respected with a tolerance of + 5 cm. They are, of course checked on a film and also immediately after each run by examining the marks of the tires on a line of talc in comparison with the ideal trajectory.

The attitudes of vehicles which form part of the test parameters are set either by replacing the shock absorbers by threaded rods if the vehicle is immobile or by modifying the suspension springs if the vehicle is mobile, in order not to disturb its trajectory.

Measurement Equipment

Tests are carried out alternately on one or other of the test tracks and the method of a mobile measurement laboratory has been selected. This is a bus which is specially equipped for the quite special function of recording all the parameters of an impact. A technique of multiplexing provides a means of obtaining more than 100 parameters of each crash.

Complete utilization also depends on wide cinematography coverage with twelve ultra-rapid cine-cameras and two reporter's cameras.

Operation of the recorders and start of the tine-cameras form an integral part of the run sequence for reasons of safety.

Equipment Specific to Accident Reconstitutions

The installations at the technical center at Lardy are the necessary basis for reconstruction of all forms of actual accidents. However, supplementary equipment specific to each case has had to be created for some of these reconstitutions.

System of Pulley Blocks Designed for Providing Different Speeds Between Two Vehicles

A system of pulley blocks combined with the possibilities offered by these launching tracks provides a means of reconstituting all possible accident configurations between two vehicles such as that illustrated in Figure 3.

Figure 3 shows diagrammatically the preparation of a collision between two vehicles, X and Y, of 'which the trajectories form an angle of 45O and of which the velocity ratio is l/2.

Accuracy of impact point is ensured because the vehicles are finally connected tot he same main cable in every case, even if auxiliary cables and carriages are used as in Figure 3.

New auxiliary cables of smaller diameter are used for every test and we have been able to verify that their elongation is negligible during the last phase, the regulation phase, of the trajectory during which the traction force is limited to preservation of velocity.

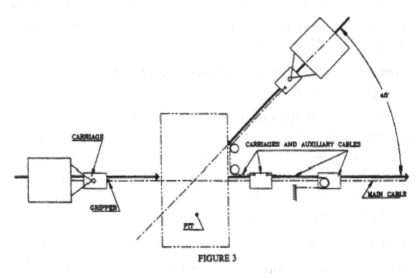

FIGURE 3

Propulsion of Pedestrians and Cyclists

A harness/electro-magnet/winch assembly maintains the pedestrian in the desired configuration while respecting position of the various centers of gravity of the subject with respect to the vehicle. A contact actuated by the vehicles controls release of the pedestrian. This contact is positioned allowing for the response time of the electro-magnet and the velocity of the vehicle so that the pedestrian is released a few milli-seconds before the impact.

The propulsion of a cyclist is a more difficult operation because this is a moving body with very unstable equilibrium. Its own velocity has, furthermore, a primordial influence on the trajectory of the subject during impact, mainly in lateral impacts which are the most frequent in this field.

The two wheels of the bicycle are guided in a steel section in the shape of an omega "ω." A third securing point is selected at the parcel holder or the saddle and its function is to stabilize the cyclist during the first acceleration phase. The whole assembly is fitted on a carriage which runs on the main rail of the track and which is stopped 2 to 3 meters before the point of impact: the cyclist continues on his impetus without any noticeable modification of his equilibrium.

Roll-Over Carriage

A traditional system of roll-over reproduces one phase normally called the perfect roll-over of which the rotational axis is the same as the longitudinal axis of the vehicle, which rarely occurs in reality. We have designed a carriage which comprises a large flat area, 4.5 m x 4.5 m, inclined to 20°, on which the vehicle may be set in any position. Thus it is possible to reproduce any form of roll-over which affect different parts of the structure.

Platform for Lateral Impacts Against Static Obstacle

This is a very flat and perfectly smooth platform which carries a vehicle which may be oriented in any direction with respect to the dynamometric pole, so that the collision against this type of obstacle is no more exclusively a frontal one. The carriage is braked at the moment of impact by distortion of two metal tubes set in the wall. The vehicle, the friction of which is lowered by the presence of a lubricant, may strike the pole laterally at any point and at any angle without any exterior interference (Figure 4).

Injection System for Recreating Arterial Pressure

The medical authorities rapidly realized the necessity for simulation of arterial pressure in order to reveal cerebral injuries and other hemorrhages, during reconstitutions of accidents with cadavers. A mixture of water, formal and china ink was selected for its good diffusion and fixation of carbon particles allows the detection of vessel ruptures. We have developed in collaboration with IRO (Institut de Recherche Orthopedique) a system that provides a means of infusing the subject just before the test and of maintaining constant pressure during the impact and a few seconds after.

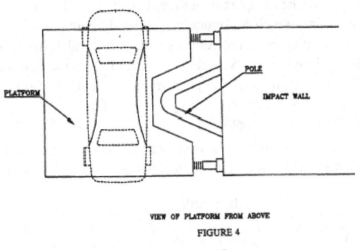

VIEW OF PLATFORM FROM ABOVE
FIGURE 4

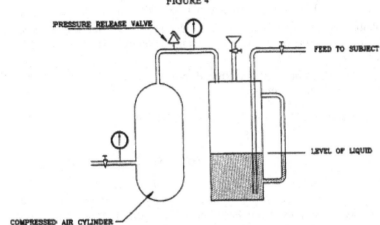

Note: This system is designed to be carried on the vehicle tested.
FIGURE 5

Conclusion

Experience acquired by more than fifty reconstitutions of actual accidents which have been carried out up to this day on the RENAULT test tracks has enabled the following conclusions to be drawn:

- If all the precautions given in detail above are taken, error of velocity is less than 1% of the value laid down and error of point of impact is less than ±5 cm.
- The other parameters which provide fidelity of the reconstitution are:

a. Accuracy of analysis of angles of impact of the true accident, which is a function of the complexity of the accident. This leads us today to prefer accidents which configurations are fairly pure.
b. Accuracy of attitudes of the vehicles and, in particular, the attitude during braking of a vehicle when colliding in the case of frontal impact, collisions with pedestrians or bicycles, and lateral collisions, or the attitude at impact time of a vehicle struck during the same lateral collisions.

All these extra precautions ensure that the deformations of vehicles are reproduced to within less than 3 cm mainly in the resistant zones of the structure.

Bibliography

Proposition d'une methode d'analyse et de classification des s&&it& de collisions en accidents reels. P. Ventre, J. Provensal - RENAULT. IRCOBI, Amsterdam 26-27 June 1973.

The contribution of physical analysis of accidents towards Interpretation of severe traffic trauma. C. Tarriere, A. Fayon, F. Hartemann, and P. Ventre - RENAULT. STAPP, San Diego, California, 17-19 November 1975.

Moyens d'essais utilises par la Regie Renault pour la n-rise au point des vehicules complets. Fontanet, Philippe, Trondle, Ventre. S.I.A., 5 May 1974.

Endnotes

1. The case citation is: Collin Johnson vs. Pedro Rivera, Peter Rivera, Nishiki of Japan, Inc., Bloornfield Bicycle, Inc., Derby Cycle Corp., Superior Court of New Jersey, Essex County, Civil Part Docket No. ESX-L-16063-89. This trial took place in the Essex County Courthouse, Newark, New Jersey in September and October, 1993.

2. Vector addition is used to show the lateral movement of the Jeep because it was too expensive to simulate the movement with the test barrier. In addition, the lateral movement of the Jeep did not contribute significant crush to the total configuration. Rather, the cycle's speed was the significant crush force. The accident site was ridden numerous times with a coasting speed of 40 mph. It is possible that the speed was higher, but 40 mph was used to be conservative. The method of vector addition employed is not perfect. However, it is a reasonable means for representing lateral speed.

3. As stated previously, this is not a perfect analysis, but a reasonable one. The cost of moving the barrier at higher speeds would be prohibitive.

4. During the Johnson trial, the French paper was finally located on microfilm (since it is no longer reprinted) and submitted to the Court. The trial judge refused to admit the paper into evidence on the grounds that it was unscientific.

5. As of 1995, the test track discussed in this chapter is still operational. Videos of the tests conducted on the track may be obtained from GE Engineering, Inc.

Chapter 23

The Civil Liability of the Forensic Engineer

Introduction

Forensic engineers frequently have questions regarding their civil liability for testimony given in connection with a legal proceeding. However, a recent Pennsylvania court decision, which is discussed below and included here in its entirety, should help answer such questions.1

In *Panitz vs. Behrend and Emsberger,* 632 A.2d 562 (Pa. Super. 1993), the court ruled that an expert witness, friendly or otherwise, is immune from civil liability for pertinent in-court testimony, as well as pertinent and material pre-trial communications made in the course of judicial proceedings. This ruling was based on the general rule that, "there is no civil liability for statements made in the pleadings or during trial or argument of a case so long as the statements are pertinent." The ruling was also based on the court's finding that, "the primary purpose of expert testimony is not to assist one party or another in winning the case but to assist the trier of the facts in understanding complicated matters."

The Panitz ruling reaffirms the established litigation or testimonial privilege. As noted by the *Panitz* court, "the purpose of the privilege is to preserve the integrity of the judicial process by encouraging full and frank testimony." The reaffirmed privilege will enable forensic engineers to act as independent professionals rather than as "hired guns."

Unfortunately, the litigation privilege may not shield the forensic engineer from specious negligence claims brought by unethical parties. All too often, disgruntled losing parties seek to cast blame on and avoid financial responsibility to their expert witnesses and others by filing meritless lawsuits. While such actions are a nuisance and financial burden, they can, in light of the above case law, likely be defeated. As the *Panitz* case reveals, the law has long expected that an expert witness will testify to the truth as he or she sees it without the constant threat of civil liability.

ELAINE B. PANITZ, M.D., Appellee v. KENNETH W. BEHREND and BARBARA BEHREND ERNSBERGER, INDIVIDUALLY and T/A BEHREND and ERNSBERGER, Appellants

No. 399 Pittsburgh, 1993

SUPERIOR COURT OF PENNSYLVANIA

632 A.2d 562; 1993 Pa. Super. LEXIS 3355

September 1, 1993, Argued
October 13, 1993, Filed

Appeal from Order of the Court of Common Pleas, Civil Division, of Allegheny County, No. 4900 of 1992. Before WETTICK, J.

DISPOSITION: Affirmed.

COUNSEL: Kenneth R. Behrend, Pittsburgh, for appellants.
Alan S. Gold, Jenkintown, for appellee.

JUDGES: BEFORE: ROWLEY, P.J., WIEAND and CIRCILLO, J.J.

OPINION BY WIEAND, J.:

Elaine B. Panitz, a medical doctor who regularly offers her services as an expert medical witness, was hired by Kenneth W. Behrend, Barbara Behrend Ernsberger and the law firm of Behrend and Ernsberger to give testimony on behalf of clients whom the law fm represented in a personal injury action. When an unfavorable verdict was returned, the lawyers refused to pay the expert witness the balance of the moneys which they allegedly had agreed to pay. Panitz sued to recover these moneys. The law firm thereupon filed an answer to the complaint which contained a counterclaim for damages resulting from the unfavorable verdict. This, it was alleged, had been caused by gross negligence and misrepresentation regarding the substance of Parritz's testimony at trial. To this counterclaim Panitz filed preliminary objections in the nature of a demurrer. When the trial court sustained the preliminary objections and dismissed the counterclaim, the defendant law firm appealed.

When reviewing an appeal from an order sustaining preliminary objections in the nature of a demurrer to a pleading, we accept as true all well-pleaded facts and all reasonable inferences to be drawn therefrom. The decisions of the trial court will be affirmed only if there is no legal theory under which a recovery can be sustained on the facts pleaded. *Allegheny County v. Commonwealth*, 507 Pa.

360, 372, 490 A.2d *402, 408* (1985); *Rutherfoord v. Presbyterian University Hospital,* 417 Pa. Super. 316, 321-322, 612 A.2d 500, 502-503 (1992).

In the underlying action, the Behrend firm had represented the Charney family whose members, allegedly, had been exposed to formaldehyde in building materials and had sustained formaldehyde sensitization reactions. Panitz was employed to support the alleged cause of action. It was expected that she would be cross-examined about the lack of such sensitization in cigarette smokers who regularly are exposed to much greater concentrations of formaldehyde than were the Charneys. In preparation for trial, Panitz provided the Behrend firm with a transcript of depositions in a prior case in which she had postulated on the lack of sensitization in smokers. Panitz testified at trial, as anticipated, that in her opinion the Charneys' injuries had been caused by formaldehyde present in building materials. When cross-examined about the lack of sensitization in cigarette smokers, however, Panitz conceded that she could not explain the apparent inconsistency.

After trial, Panitz explained that she had come to realize prior to trial that the reasoning upon which she had relied in earlier depositions was inaccurate. As a general rule there is no civil liability for statements made in the pleadings or during trial or argument of a case so long as the statements are pertinent. *Post v. Mendel, 5* 10 Pa. 213,221,507 A.2d 35 1,355 (1986); *Greenberg v. Aetna Ins.* Co., 427 Pa. 5 11,5 16,235 A.2d 576,577 (1967), cert. denied, *Scarselletti x Aetna Ins.* Co., 392 U.S. 907, 88 S.Ct. 2063, 20 L.Ed.2d 1366 (1968); Moses v. *McWilliams,* 379 Pa. Super. 150, 163, 549 A.2d 950, 956 (1988), allocatur denied, 521 Pa. 631,558 A.2d *532* (1989); *Pelagatti* V. *Cohen,* 370 Pa. Super. 422, 436, 536 A.2d 1337, 1344 (1987), allocatur denied, 519 Pa. 667, 548 A.2d 256 (1988). The privilege, which includes judges, lawyers, litigants and witnesses, had its origin in defamation actions premised upon statements made during legal actions, but it has now been extended to include all tort actions based on statements made during judicial proceedings. Thus, in *Clodgo by Clodgo v. Bowman,* 411 Pa. Super. 267, 601 A.2d 342 (1992), the judicial or testimonial privilege was held to insulate a court appointed medical expert witness from liability premised upon malpractice. See also: *Moses v. McWilliams, supra; Brown v. Delaware Valley Transplant Program,* 371 Pa. Super. 583,538 A.2d 889 (1988). "The form of the cause of action is not relevant to application of the privilege. Regardless of the tort contained in the complaint, if the communication was made in connection with a judicial proceeding and was material and relevant to it, the privilege applies." *Clodgo by Clodgo v. Bowman,* supra at 273,601 A.2d at 345. The privilege is equally applicable where the cause of action is stated in terms of misrepresentation or a contractual requirement to exercise due care.

The purpose for the privilege is to preserve the integrity of the judicial process by encouraging full and frank testimony. This was recognized by the Supreme Court of the United States in *Briscoe v. LaHue,* 460 U.S. 325, 103 S.Ct. 1108, 75 L.Ed.2d 96 (1983), where the Court said:

"The claims of the individual must yield to the dictates of public policy, which requires that the paths which lead to the ascertainment of truth should be left as free and unobstructed as possible." *Calkins v. Sumner;* 13 Wis. 193, 197 (186). A witness' apprehension of subsequent damages liability might induce two forms of self-censorship. First, witnesses might be reluctant to come forward to testify. See *Henderson v. Broomhead,* supra, 578-579, 157 Eng. Rep., at 968. And once a witness is on the stand, his testimony might be distorted by the fear of subsequent liability. See *Barnes v. McCrate,* 32 Me. 442, 446447 (185 1) . . . A witness who knows that he might be forced to defend a subsequent lawsuit, and perhaps to pay damages, might be inclined to shade his testimony in favor of the potential plaintiff, to magnify uncertainties, and thus to deprive the finder of fact of candid, objective, and undistorted evidence. See Veeder, *Absolute Immunity in Defamation: Judicial Proceedings,* 9 Colum.L.Rev. 463, 470 (1909).

Id. at 332-333, 75 L.Ed.2d at 103-104, 103 S.Ct. at 1114. Similarly, the Supreme Court of Pennsylvania, in Binder v. Triangle Publications, Inc., 442 Pa. 319, 275 A.2d 53 (1971), explained:

The reasons for the absolute privilege are well recognized. A judge must be free to administer the law without fear of consequences. This independence would be impaired were he to be in daily apprehension of defamation suits. The privilege is also extended to parties to afford freedom of access to the courts, to witnesses to encourage their complete and unintimidated testimony in court, and to counsel to enable him to best represent his client's interests.

Binder v. Triangle Publications, Inc., supra at 323-324, 275 A.2d at 56. See also: *Moses v. McWilliams,* supra at 164, 549 A.2d at 957. The privilege, thus, serves the salutary purpose of encouraging witnesses to give frank and truthful testimony. Having testified truthfully in the judicial process, a witness should not thereafter be subjected to civil liability for the testimony which he or she has given.

"Witness immunity should and does extend to pre-trial communications. The policy of providing for reasonably unobstructed access to the relevant facts is no less compelling at the pre-trial stage of judicial proceedings." *Moses v. & McWilliams,* supra at 166, 549 A.2d at 958. Thus, the privilege is not to be avoided by the disingenuous argument that it was not the in-court testimony that caused the loss but the pre-trial representations about what the in-court testimony would be. The privilege includes all communications "issued in the regular course of judicial proceedings and which are pertinent and material to the redress or relief sought." *Post* V. Mended, supra at 221,507 A.2d at 355 (emphasis omitted). *See also: Kahn v. Burman,* 673 F. Supp. 210,213 (E.D. Mich. 1987), aff'd, 878 *E 2d 1436 (6th* Cir. 1989); *Greenberg v. Aetna Ins. Co.,* supra at 516, *235 A.2d* at 5'7'7 (applying privilege to statements in pleadings); *Moses v. McWilliams,* supra (applying privilege to statements made by witness to attorney during pre-trial *conference); Pehgatti v. Cohen,* supra at 436,536 A.2d at 1344 (applying privilege to statements in pre-trial affidavits). The "expert's courtroom testimony is the last act in a long, complex process of evaluation and consultation with the litigant. There is no way to distinguish the testimony from the acts and communications on which it is based." *Bruce v. Byrne-Stevens & Associates Engineers,* Inc., 113 Wash. 2d 123, 135, 776 P.2d 666, 672 (1989). See also: *Middlesex Concrete Products v. Carteret Ind. Ass'n,* 68 N.J. 85,92, 172 A.2d 22,25 (1961).

There also is no reason for refusing to apply the privilege to friendly experts hired by a party. The policy of encouraging frank and objective testimony, without ear of civil liability therefor, "obtains irrespective of the manner by which the witness comes to *court." Bruce v. Byrne-Stevens & Associates Engineers, Inc.,* supra at 129, 776 I?2d at 669. The primary purpose of expert testimony is not to assist one party or another in winning the case but to assist the trier of the facts in understanding complicated matters.

In *Bruce v. Byrne-Stevens & Associates Engineers, Inc.,* supra, a claim was made against an engineering witness on grounds that the expert had negligently miscalculated the cost of restoring lateral support to plaintiff's land. In dismissing the action, the Supreme Court of Washington held that (1) an expert witness was entitled to the privilege even if he or she had been retained and compensated by a party rather than appointed by the court, and (2) immunity extended not only to in-court testimony but also to acts and communications which had occurred in connection with preparing that testimony. The court reasoned:

While it may be that many expert witnesses are retained with the expectation that they will perform as "hired guns" for their employer, as a matter of law the expert serves the court. The admissibility and scope of the expert's testimony is a matter within the court's discretion. *Orion Corp. v. State,* 103

Wash.2d 441, 462, 693 P.2d 1369 (1985). That admissibility turns primarily on whether the expert's testimony will be of assistance to the finder of fact. ER 702. The court retains the discretion to question expert witnesses. ER 614(b). The mere fact that the expert is retained and compensated by a party does not change the fact that, as a witness, he is a participant in a judicial proceeding.

Id. at 129-130, 776 P.2d at 669. A contrary holding, the court concluded, would result in a loss of objectivity in expert testimony generally and would discourage anyone who was not a professional witness from testifying. Although an intermediate appellate court in California has held that the privilege should not apply to a "friendly" expert, see: *Mattco Forge, Inc. v. Arthur Young & Co.,* 5 Cal. App.4th 392,6 Cal.Rptr. 78 1 (1992), and an intermediate appellate court in Arizona has implied in dictum that it would not be inclined to extend the privilege to "friendly" experts, see: *Lavit v. Superior Court,* 173 Ariz. 96, 839 P.2d 1141 (Ariz. Ct. App. 1992), we conclude that the better view is that followed by the Supreme Court of Washington. To allow a party to litigation to contract with an expert witness and thereby obligate the witness to testify only in a manner favorable to the party, on threat of civil liability, would be contrary to public policy. *Griffith v. Harris,* 17 Wis.2d 255, 116 N.W.2d 133 (1962), cert. denied, 373 U.S. 927, 83 S.Ct. 1530, 10 L.Ed.2d 425 (1963). See also: *Curtis v. Wolfe,* 160 Ill. App. 3d 588,513 N.E.2d 1139, 112 Ill. Dec. 530 (1987), appeal denied, 118 Ill. 2d 542, 520 N.E.2d 384, 117 Ill. Dec. 223 (1988). "Fundamentally, no witness can be required to testify, and no witness should be expected to testify, to anything other than the truth as he [or she] sees it and according to what he [or she] believes it to be. The same is expected of expert witnesses." *Schafer v. Donegan,* 66 Ohio App. 3d 528,538,585 N.E.2d 854,860 (1990), jurisdictional motion overruled, 55 Ohio St.3d 722,564 N.E.2d 500 (1990). An expert witness will not be subjected to civil liability because he or she, in the face of conflicting evidence or during rigorous cross-examination, is persuaded that some or all of his or her opinion testimony has been inaccurate.

Because appellee was immune from civil liability for testimony which she gave, the trial court properly dismissed the counterclaim against her.

Affirmed.

Endnotes

1. It should be noted that the opinions discussed in this paper are not legal opinions. Rather, they are based on an analysis of the subject case law from the perspective of the forensic engineer. It is, therefore, recommended that the forensic engineer seek proper legal advice before making any determination as to his or her individual civil liability.

Chapter 24

An Analysis of the Force Required to Release Cycling Shoes from Clipless Pedals

Introduction

During the first 90 years of competitive and recreational cycling, cyclists strapped their feet to pedals using straps threaded through the pedals and wrapped around the feet. The pedal/foot strapping afforded cyclists a method for lifting the pedal while riding. As the pedal design process progressed, the cycling industry developed a toe clip device with a strap already threaded through it. This device required the cyclist to insert the foot into the toe clip, or cage as it was sometimes called, and then reach down and fasten the foot securely to the pedal by pulling the toe strap tight.

The advantage of the aforementioned toe clip device was that it allowed the rider to impact as much as 30 percent more force to the pedaling motion. The disadvantage of the device was that it required the cyclist to reach down and loosen the strap prior to removing the foot. Failure to loosen the strap would result in the cyclist falling over when coming to a stop. The actual force required to release the foot from the cage type of toe clip has never been studied. However, without releasing the toe strap, testing has shown that the force required for release is so high that destruction of the shoe and toe clip results. This is at least 50 pounds of peak force applied perpendicular to the side of the shoe. It can thus be safely surmised that, for all practical purposes, the old type of shoe and cage toe clip arrangement does not release without loosening the strap.

In the early 1980's, a French company named Look developed a new pedal design very similar to a downhill ski binding. The new design enables the cyclist to step into the pedal. The cycling shoe attaches to the pedal with a specially designed cleat on the bottom of the shoe. An audible click can be heard when the cleat properly engages the pedal. The cyclist's foot will only release from the pedal when lateral motion is applied. This means that the cyclist must force his/her foot away from the bike by applying pressure to the inside or outside rear of the cycling shoe. Consequently, the cyclist does not have to reach down and unfasten a strap, but simply pushes his/her foot away from the bike to obtain the release.

Despite the advantages of the new cleat pedal system, several allegations have been made since its introduction regarding the difficulty of removing the foot from the system. These allegations have arisen in connection with claims that the pedal system was a causal factor in cycling accidents. In light of these allegations, the purpose of this analysis is to determine whether there are any appreciable differences between the different manufacturers' shoes and to define the different release rates.

Engineering Analysis

In order to determine the lateral release values of the various cleat cycling shoes, a clipless pedal test jig was constructed, Figure 1. Each shoe was mounted on the appropriate pedal, which was supplied by the manufacturer. Shoes and pedals were matched in accordance with the accompanying manufacturer's literature. If any cleat or pedal adjustments were needed, the lowest release value was always selected. Photographs of the test jig are shown in Figures 2 and 3.

A stainless steel probe was fastened to the side of a shoe by a metal plate. The plate was positioned so that no significant play was apparent in the side of the shoe when lateral force was applied to the system. A 100 pound gauge was used to measure the lateral force placed on the shoe by the probe. The force was applied using a crosshead and screw system. The speed of the screw was measured and controlled by a digital gauge and motorized arrangement. In effect, the system was able to accurately measure the lateral force that a cyclist would place on the right side rear of the shoe to release the foot from the pedal.

Analytical Results

The amount of lateral displacement required to measure the peak force needed for shoe release is noted in Table 1. Lateral displacement is the distance the shoe should move in order to achieve release. The peak force, in pounds, is the amount of force required to achieve the necessary lateral displacement.

The pedal system with the minimum peak force requirement for release was the Mavic 646LMS with 10.4 pounds. This system also had the minimum displacement istance of 1.09 inches. The Mavic 646LMS pedal release system was the most efficient release system of those studied.

In contrast, the Wellgo RG9203 pedal system did not release adequately. The shoe tended to catch on the pedal and not release from lateral force application. It was not possible to obtain a lateral force value on this system.

Similarly, the Speed Play Bryne Xl2 needed a peak force outside the limits of the system. The shoe releases from the pedal outside the 100 pound measuring limit of the testing jig.

The Campagnolo Record QR released cleanly at 1.23 inches of displacement and 15.3 pounds of force. However, this system allowed the shoe cleat to catch a second time on the pedal.

The other systems studied all released cleanly at the displacement and peak force values noted in Table 1.

Conclusion

With the exception of the three clipless pedal systems noted above, the clipless pedal manufacturers have a very safe product.' Indeed, the low displacement and peak force values revealed in this analysis establish the implausibility of most allegations that the clipless pedal systems are accident causal factors.

Given the low lateral release values of these systems, a cyclist's foot should disengage from the pedal in a fall from the cycle. The low values also allow a cyclist to easily disengage his/her foot when coming to a stop. The foot and leg are capable of easily placing close to 50 pounds of force laterally on the shoe. Thus, the lateral peak force values of 10 to 15 pounds are easily obtained by even the weakest rider.

A comparison of various clipless pedal systems is noted in Figure 4. The forensic engineer should compare the pedal system under investigation to this Figure. It is also recommended that the subject pedal system be tested in the laboratory. Please note that this analysis is not exhaustive. Issues not addressed here include use of wrong pedal with wrong cleat; improper assembly of the clipless pedal system; and misadjustment of the pedal adjustment screws. These issues should, however, be examined during any accident reconstruction involving a clipless pedal system.

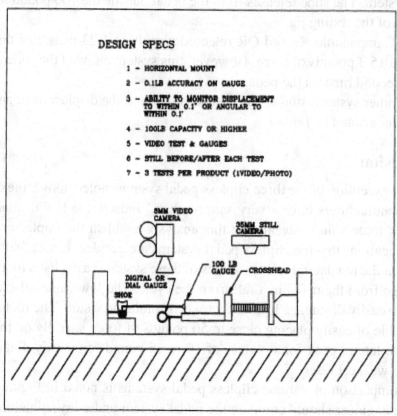

Figure 1 Clipless Pedal Test Jig

Figure 2 Overview of Testing Jig

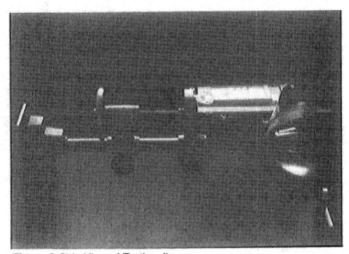
Figure 3 Side View of Testing Jig

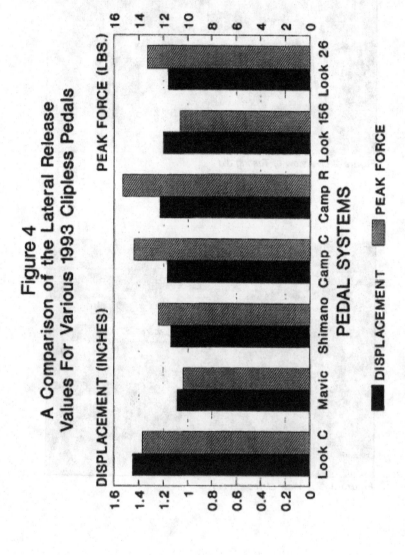

Figure 4
A Comparison of the Lateral Release Values For Various 1993 Clipless Pedals

Pedals capable of purchase in 1993

TABLE 1

A COMPARISON OF THE LATERAL RELEASE VALUES FOR VARIOUS 1993 CLIPLESS PEDALS

PEDAL		DISPLACEMENT (inches)	PEAK FORCE (pounds)
Look Carbon Pro	x_1	1.44	13.8
	x_2	1.44	13.7
	x_3	1.47	13.7
	Average $\bar{x}_3 =$	1.45	13.7
Mavic 646LMS	x_1	1.09	10.4
	x_2	1.09	10.5
	x_3	1.09	10.3
	Average $\bar{x}_3 =$	1.09	10.4
Shimano PD-7401	x_1	1.16	12.6
	x_2	1.13	12.3
	x_3	1.13	12.4
	Average $\bar{x}_3 =$	1.14	12.4
Campagnolo Chorus QR	x_1	1.19	14.4
	x_2	1.16	14.4
	x_3	1.16	14.3
	Average $\bar{x}_3 =$	1.17	14.4
Campagnolo Record QR	x_1	1.28	15.5
	x_2	1.28	15.3
	x_3	1.19	15.3
	x_4	1.19	15.2
	x_5	1.22	15.2
	Average $\bar{x}_3 =$	1.23	15.3
Look PP156	x_1	1.16	10.6
	x_2	1.19	10.6
	x_3	1.25	10.7
	x_4	1.19	10.7
	Average $\bar{x}_3 =$	1.20	10.65

Look PS26	x_1	1.16	13.8
	x_2	1.16	13.3
	x_3	1.13	13.0
	x_4	<u>1.19</u>	<u>13.1</u>
	Average $\bar{x}_3 =$	1.16	13.3

Wellgo RG9203 — Shoe had multistaged release and could not be measured. Shoe tended to catch on pedal and not release from lateral force application.

Speed Play Bryne x/2 — The peak force needed was outside the limits of the system. Shoe will not release with 100 pounds of lateral force.

Endnotes

1. The Campagnolo Record QR has been retrofitted to correct this catching problem. As of 1994, the Campagnolo system is within the ranges noted in Figure 2.

Chapter 25

The Use of EDSMAC to Simulate Bicycle And Motor Vehicle Collisions

Introduction

The purpose of this chapter is to illustrate how licensed users of the EDSMAC program can simulate motor vehicle/bicycle collisions.

EDSMAC is an automobile collision simulation model. Specifically, the model simulates vehicle dynamics at impact and velocity change as a result of impact. The model is commonly accepted as authoritative by forensic engineers and courts. Thus, the revised input values provided here should successfully aid in the production of acceptable and reliable motor vehicle/bicycle collision simulations.

The revised input values have been verified in the field. Applicable field data is included. Also included is an actual EDSMAC computer run. Analysis of the computer run focuses on the kinematics of a motor vehicle/bicycle collision up to the point of impact ("POI"). The vehicle and bicycle speeds achieved during the run were verified in the field.

Post impact kinematics are more difficult to determine with the modified EDSMAC program. However, my field investigations indicate that the program can closely approximate the post impact kinematics of a cycle for a period of at least 0.5 seconds post impact, if the cycle stays on the ground. Work is continuing on this area of the program. It must also be noted that the revised input values deliberately introduce error codes and default values into the EDSMAC program in order to initialize the modified program. The error codes and default values do not, however, impact a computer run of a motor vehicle/bicycle collision simulation.

Although the modified EDSMAC program has some limitations, field verification has shown that it will produce very accurate results up to the point of impact.

Input Changes to EDSMAC

The key bicycle inputs for EDSMAC are noted under Vehicle Dimensional and Inertial Properties. Vehicle 1 was chosen to be the cycle. Wheelbase is noted as 48.0 inches from an exemplar cycle. The wheelbase is measured from the point of contact of the front and rear wheels with the ground. Whenever practicable, wheelbase should be measured from the actual cycle. The applicable center of gravity (Cg) is the combined mass of the bicycle, the cyclist, and any normally attached gear (such as panniers, handlebar bags, carriers, etc.). The Cg is expressed as weight and is assumed to be 15 pounds for this application. The location of the Cg in relation to the front and rear axle is 0.5 * wheelbase or 24 inches in this example. The overall length of the cycle is the distance from plumb lines tangential to the front and rear wheel. The overall length in this example is 70 inches. Therefore, the Cg to both the front and rear end is 0.5 * overall length or 35 inches. The vehicle 2 parameters are for an exemplar motor vehicle chosen from the EDSMAC data bank. YAW moment of inertia is a default value which will have no impact on the EDSMAC simulation. All values noted assume a cyclist in a seated position on level ground. Example calculations are shown as follows:

Example Calculations for Vehicle Dimensional and Inertial Properties for Simulating a Bicycle

Parameter	Calculation or Measurement	Input Value
Vehicle Class Category	Use 1 to run program	1
Wheelbase	Measured Value from Cycle	48.0 inches
Cg to Front Axle	0.5 * 48 inches	24.0 inches
Cg to Rear Axle	0.5 * 48 inches	24.0 inches
Truck Width	Width of tire on ground	1.0 inches
Overall Length	Measured value from cycle	70.0 inches
Cg to Front End	0.5 * 70 inches	35.0 inches
Cg to Rear End	0.5 * 70 inches	35.0 inches
Overall Width	Measured width of tire & rim	2.0 inches
Weight	Total weight of rider cycle system	150 pounds

The above calculations are only examples. They are not from an actual reconstruction.

For the example simulation, Vehicle 1 was chosen to be the cycle. The settings given were as follows:

Initial Conditions

	Vehicle 1	Vehicle 2
X - Coordinate	47.0 ft	0.00 ft
Y - Coordinate	-1.00 ft	37.0 ft
PSI - Coordinate (heading)	180 degrees	270 degrees
U - Velocity	10.0 mph	10.0 mph
V - Velocity	0.0 mph	0.0 mph
YAW Velocity	0.0 mph	0.0 mph

The motor vehicle, Vehicle 2, was given a surprise reaction steer angle of 8 degrees to be applied 2 seconds into the simulation.

The input for the exemplar motor vehicle becomes:

Wheel Force and Steer Angle - Vehicle No. 2 (Motor Vehicle)

Wheel Force - Pounds				Time (sec)	Steer Angle (deg.)	
R/F	L/F	R/R	L/R		R/F	L/F
-0.01	-0.01	0.01	0.01	0.000	0.00	0.00
				2.000	8.00	8.00

For the example simulation, the cycle was assumed to proceed in a straight path. But the above steer angles can be applied to the cycle in other simulations. Cycle steer angle should, however, be determined in the field using the accident cycle or an exemplar cycle.

The input for the exemplar bicycle becomes:

Wheel Force and Steer Angle - Vehicle No. 1 (Bicycle)

Wheel Force - Pounds				Time (sec)
R/F	L/F	R/R	L/R	
-0.01	-0.01	0.01	0.01	0.000

No steer angle was utilized.

Output of EDSMAC Trajectory Simulation

The output of the example EDSMAC simulation is shown in Figure 1. The cycle approaches along the X-axis at 10 mph. The motor vehicle approaches at an 8 degree steer angle which is executed 2 seconds into the scenario. The point of impact on the motor vehicle occurs approximately one-third of the way back from the front of the motor vehicle. The cycle is moved laterally in the direction of the car's path of travel. This movement leaves skid marks on the pavement. A coefficient of friction of 0.700 is assumed as an input value. This is noted as nominal road friction. The EDSMAC program is actually capable of showing skid marks from the cycle even though the mass is reduced from that of a motor vehicle to that of a cycle/rider system.

Conclusion of the Analysis Using EDSMAC to Simulate Bicycle and Motor Vehicle Collisions

Field tests were conducted to validate EDSMAC's ability to simulate the travel time of a bicycle. Four different cycles were used in the tests. Each cycle's Time of Travel (TOT) over a calibrated course was compared to the TOT or Velocity predicted in EDSMAC. The results of the field tests are noted in Table 1.

Table 1
Predicted Time of Travel from EDSMAC Compared to Field Measured Time of Travel on a 100 ft. Calibrated Course

Bicycle	5 mph	10 mph	15 mph	20 mph
Kestrel 10-speed	13.61	6.79	4.49	3.42
Raleigh SP1200	13.53	6.80	4.56	3.41
De Rosa 10-speed	13.58	6.83	4.48	3.42
Haro 24-speed	13.63	6.79	4.50	3.39
EDSMAC	13.61	6.80	4.54	3.40

The above data reveals the following:

1. All cycles were ridden by human subjects at rpm's sufficient to maintain the required velocity on a calibrated velometer. TOT was then measured using a Robic SC-600 chronometer with 1/100th precision. The results are also plotted in Figure 1.
2. EDSMAC is an extremely accurate program for plotting the approach speed of a cycle up to POI. The data plotted in Figure 1 and tabulated in Table 1, reveals no significant difference between field values and EDSMAC values for a cycle's TOT.
3. Figure 2 shows the Trajectory Plot of the exemplar collision. It should be noted that EDSMAC produces reasonably accurate estimates of cycle dynamics if it is assumed that the exemplar cycle stays on the ground. Also, the forensic engineer will need to use the trajectory calculations discussed elsewhere in this book for simulations of both the cyclist and the cycle.
4. The modified EDSMAC program discussed in this chapter accu-

rately simulates motor vehicle/bicycle collisions up to POI. The program is also a good means for developing an index for immediate post-impact collisions, if the cycle stays on the ground. Unfortunately, the modified program cannot be used for computing the trajectory of the cyclist after impact.

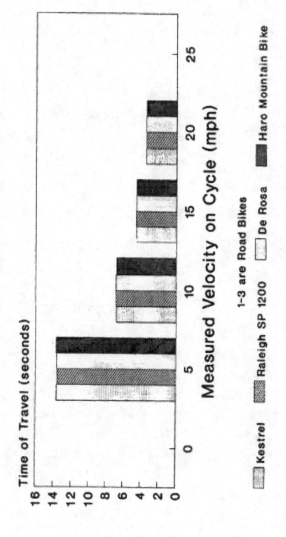

Figure 1
Predicted Time of Travel from EDSMAC
Compared to Field Measured Values

EDSMAC Results are; 5mph = 13.61,
10mph = 6.80, 15mph = 4.54,
20mph = 3.40 (Results in Seconds)

The Use of EDSMAC to Simulate Collisions 209

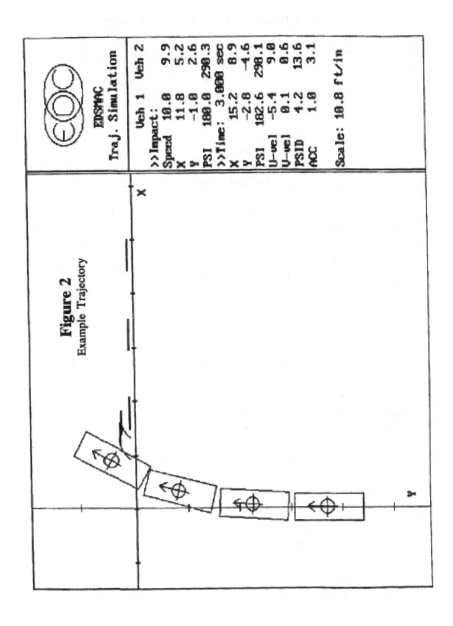

Figure 2
Example Trajectory

Summary of EDSMAC Results

Example Reconstruction - Cycle at 10 mph

At the time execution was halted, both vehicles were still moving with the following velocities:

Vehicle No.	Linear Vel. ft/sec	Angular Vel. deg/sec
1	8.0	4.2
2	13.2	13.6

Accident History

	time	POSITION			VELOCITY		
		X	Y	psi	U	V	angular
Beginning of Simulation							
Veh #1	0.00	47.0	-1.0	180.0	10.0	0.0	0.0
Veh #2		0.0	37.0	270.0	10.0	0.0	0.0
Impact							
Veh #1	2.40	11.8	-1.0	180.0	10.0	0.0	0.0
Veh #2		5.2	2.6	290.3	9.9	0.6	14.9
Separation							
Veh #1	2.60	12.0	-1.8	180.5	-5.6	4.5	5.3
Veh #2		6.4	0.1	292.6	9.0	0.5	13.5
End of Simulation							
Veh #1	3.00	15.2	-2.8	182.6	-5.4	0.1	4.2
Veh #2	3.00	8.9	-4.6	298.1	9.0	0.6	13.6

Program Control Data

Initial Simulation Time	0.000 sec **
Maximum Simulation Time	3.000 sec
Collision Phase Integration Time Interval	0.010 sec
Separation Phase Integration Time Interval	0.010 sec **
Trajectory Phase Integration Time Interval	0.050 sec **
Output Print Interval	0.050 sec **
Minimum Linear Velocity for Stopping Test	2.000 mph **
Minimum Angular Velocity for Stopping Test	5.000 deg/sec **

Initial Conditions

	Vehicle #1	Vehicle #2
X-Coordinate	47.00 ft	0.00 ft
Y-Coordinate	-1.00 ft	37.00 ft
PSI-Coordinate (heading)	180.00 deg	270.00 deg
U-Velocity	10.00 mph	10.00 mph
V-Velocity	0.00 mph	0.00 mph
YAW Velocity	0.00 deg/sec	0.00 deg/sec

Vehicle Dimensional and Inertial Properties

	Vehicle #1	Vehicle #2
Vehicle Class Category	1	1
Wheelbase	48.00 in	93.20 in **
CG to Front Axle	24.00 in	45.10 in
CG to Rear Axle	24.00 in	48.10 in **
Track Width	1.00 in	51.10 in **
Overall Length	70.00 in	159.80 in **
CG to Front End	35.00 in	76.00 in **
CG to Rear End	35.00 in	83.80 in **
Overall Width	2.00 in	60.80 in **
Rear Axle Steer Angle	0.00 deg **	0.00 deg **
Weight	150.00 lb	2202.00 lb **
YAW Moment of Inertia	11432.00 lb-sec^2-in **	11432.00 lb-sec^2-in **

** indicates default value

Wheel Force and Steer Angle Tables
Vehicle No. 1

Time (sec)	Wheel Force (lb)				Time (sec)	Steer Angle (deg)	
	R/F	L/F	R/R	L/R		R/F	L/G
0.000	-0.01	-0.01	0.01	0.01			

Wheel Force and Steer Angle Tables
Vehicle No. 2

Time (sec)	Wheel Force (lb)				Time (sec)	Steer Angle (deg)	
	R/F	L/F	R/R	L/R		R/F	L/G
0.000	–0.01	–0.01	0.01	0.01	0.000	0.00	0.00
					2.000	8.00	8.00

Endnotes

1. EDSMAC is a licensed program available from Engineering Dynamics Corporation in Beaverton, Oregon. Please note that standard EDSMAC procedures and use are not discussed here since classes on those matters are offered to licensed users.

Chapter 26

The Role of Alcohol in Reconstructing Bicycle Accidents

Introduction

The use of alcohol by cyclists has been virtually ignored in the engineering and scientific literature utilized in bicycle accident reconstructions. However, the forensic engineer has a substantial amount of literature available for the reconstruction of motor vehicle accidents involving alcohol. In fact, the laws in most states limiting the legal blood alcohol content (BAC) of drivers are based on exhaustive studies of the human physiology's tolerance to alcohol. Also, forensic engineers have readily utilized breathalyzer test results to define the point at which the physical impairment of the driver is a contributing factor to the accident. The purpose of this analysis is to utilize the existing scientific and technical literature to define the point at which alcohol ingestion is a contributing factor to a bicycle accident.

Analysis of the Literature

The correlation between bicycle accidents and the consumption of alcohol was documented in a landmark study conducted by Guohua Li and Susan P. Baker at John Hopkins University (Li study).[1] Prior to the Li study, alcohol involvement in bicycling injury was not rigorously documented in the available literature.

The Li study essentially documented what police officers and forensic engineers already suspected. That is, there is a correlation between fatally injured bicyclists aged 15 years and older and BAC levels. For example, the study showed that of 1,711 deceased bicyclists aged 15 or older who were tested for alcohol, 32% had positive BAC's and 23% were legally intoxicated. Even among victims ages 15 to 19 who are legally prohibited from drinking, 14% had positive BAC's.

As expected, the Li study also revealed that cyclist BAC levels were higher in fatal nighttime accidents. Of 877 cyclists ages 15 and older who were fatally injured at night, 30.9% had BAC's greater than 0.10%. This figure is significantly higher than the 13.4% of cyclists (754 in same age group) who were fatally injured during daylight hours. The fact that potentially one-third of all cyclists fatally injured in nighttime accidents may be legally intoxicated presents an additional area of investigation for the forensic engineer.

Although the Li study admits to some differences in population sampling between states, it is probably safe to assume that approximately 28% of cyclist fatalities have a BAC of greater than 0.10%. Yet, most forensic engineers fail to consider the use of alcohol when determining reaction time and time of travel during bicycle and pedestrian accident reconstructions. Such omissions may likely be the result of the previously mentioned dearth of relevant literature. To avoid similar omissions, the forensic engineer should, when applicable, utilize the calculations discussed below.

Factoring in Alcohol Consumption In Bicycle and Pedestrian Accident Reconstructions

If the forensic engineer is aware that a cyclist has a BAC greater than 0.10%, he/she must determine the pharmacological and psychomotor aspects of the alcohol consumed for a thorough and accurate reconstruction of the accident. Until recently, this determination could not be made without an exhaustive search of the literature. However, the work of James Garriott has brought together all of the pertinent literature for the forensic engineer[2].

As Garriott's work suggests, where cyclist alcohol consumption is involved, the primary consideration in the accident reconstruction is the reaction time available to the cyclist. To quantitatively define the increase in reaction time, th results of a standard breathalyzer test are used to give the following relationship, as defined by the Uniform Motor Vehicle Code. The amount of alcohol in the deep lung (DL) is equivalent to the amount of alcohol in the blood supply to the brain. Therefore, alcohol concentration shall mean either grams of alcohol per 100 milliliters of blood or grams of alcohol per 210 liters of breath. Thus, a typical conversion would be:

$$0.15\% \text{ wt./vol.} = .15 \text{ g/l or } 150 \text{ mg/dl}$$

If the forensic engineer prefers to work from the initial breathalyzer analysis, the conversion is to liters, as follows:

Given that the breathalyzer test is 3 1,500 mg/210 l, or mg of alcohol as ethanol per liters of air.

$$\text{Then; } \frac{31,500 \text{ mg}}{210 \text{ l}} = 150 \text{ mg/l or } .15 \text{ g/l or } 0.15\% \text{ wt/vol},$$

where the wt/vol is the BAC.

The breathalyzer test is, of course, delivered to live subjects.

If the cyclist suffered a fatality, the forensic engineer will have to rely on the coroner's report for BAC **immediately at the time of death.** If the BAC is not listed in the coroner's report, the forensic engineer should ascertain whether specimens were taken and analyzed by emergency room personnel or the coroner which might reveal the necessary data.

Fortunately, alcohol distribution in the body has been determined quantitatively since as early as 1937[3].

TABLE 1

Alcohol Distribution Ratios in Various Organs and Body Fluids After Equilibrium

	Distribution Ratio	Water Content (%)
Whole Blood	1.00	78.6
Brain - Blood	0.847	76.3
Liver - Blood	0.797	73.2
Femoral Muscle - Blood	0.856	70.6
Stomach Contents - Blood	0.958	92.5
Stomach Tissue - Blood	0.788	79.9
Upper Intestine - Blood	0.788	55.7
Lower Intestine - Blood	0.876	77.4
CSF - Blood	1.180	92.7

The distribution ratios noted in Table 1 are a critical element of the accident reconstruction where the cyclist or pedestrian is "broken up" from the impact with a motor vehicle.

Case Study

A cyclist was struck and killed by a hit and run driver at 3:00 a.m. in an intersection of a large city. The plaintiffs contended that the cycle was defective because the bike company did not attach a headlight to the bicycle and this defect caused the accident.

The cyclist was "broken up" in multiple areas from the impact. Fortunately, the coroner performed an ethyl alcohol analysis on the cyclist's liver. Thus, the following calculation was used:

Given that the liver blood had 198 mg/l of ethanol, the BAC is:

$$198 \text{ mg/l} = .198 \text{ g/l or } 0.198\% \text{ BAC of the liver}$$

To calculate the BAC of the circulatory system, we use Table 1, where:

$$\frac{1.00}{0.797} = \frac{x}{0.198\%}$$

$$0.797 x = (1.00)(0.198\%)$$

$$x = \frac{0.198\%}{0.797} = 0.248\% \text{ BAC of the circulatory system.}$$

An investigation of the cyclist's activities during the day of the accident revealed that he had been drinking steadily in area bars because he had a fight with his girlfriend. This fact was never revealed at trial since the case immediately settled.

Summary

The work of Li, Garriott and others clearly shows that individuals with a BAC greater than 0.10% cannot safely and effectively operate vehicles, including bicycles, because they are physically impaired. Despite this fact, a surprisingly large segment of the cycling population attempts to ride with a BAC greater than 0.10%. Consequently, alcohol consumption is a significant causal factor of cycling accidents.

The work of Li and Garriott has changed the way the forensic engineer should approach investigations of bicycle and pedestrian accidents involving alcohol consumption. These changes are as follows:

1. One of the principal facts that should be ascertained is the cyclist's BAC. As noted in the Li study, even cyclists aged twelve and under may have a BAC greater than 0.10%. Note that the BAC levels of both the motorist and cyclist are often taken by the investigating police officer. However, the results are not always put on the accident report[4].
2. Time weighted BAC values must be carefully evaluated if the cyclist was fatally injured in the accident. A physician is absolutely essential to aid in back calculating the BAC at the time of death. The methods for determining BAC that are presented in this chapter have been accepted by the engineering and scientific community and should be admissible in court.

Endnotes

1. Li, Guohua and Baker, Susan P., *"Alcohol in Fatally Injured Cyclists,"* Injury Prevention Center, The John Hopkins University School of Hygiene and Public Health, Baltimore, Maryland. Presented at the 37th Annual Proceedings of the Association for the Advancement of Automotive Medicine, November 4-6, 1993, San Antonio, Texas.
2. Garriott, James C. *"Medicolegal Aspects of Alcohol Determination in Biological Specimens,"* Lawyers and Judges Publishing Company, 1993.
3. Harger, R.N., Hulpieu, R.H., and Lamb, E.B., *"The Speed With Which Various Parts of the Body Reach Equilibrium in the Storage of Ethyl Akohol."* J. Biol. Chem. 120:689-704, 1937. Review of this data indicates that the results with dogs correspond very closely with the results achieved with humans. If the accident reconstruction warrants the cost, simulations could be done with dogs. Live of course!
4. I once had an investigation where the fact that the investigating officer took the BAC of both the motorist and the cyclist was not known until trial, when the officer released his complete file to the court.

Chapter 27

An Evaluation of a Signed Release in Reconstructing Bicycle Accidents

Introduction

Before participating in a bicycle race or triathlon, a competitor is often asked to sign a waiver or release form. The form usually provides that the competitor agrees to "hold harmless" the race directors in the event of an accident. Release forms are widely used by race administrators because of the significant risk of injury to a race competitor.

Bicycle races may be, and typically are, mass start events that are run point to point or on a closed circuit. Bicycle races may also be time trials in which competitors race alone, from point to point, against the clock. Triathlons are swim, bike, and run competitions with a mass start at the beginning of the swim. Competitors race against the clock and each other during the bike and run legs. In a biathlon, which consists of a bike and run, the mass start is typically at the run. [The governing body of cycle racing is the United States Cycling Federation ("USCF").] During these racing events, competitors may be injured in the "crush" of the mass start or by unforeseen and foreseen hazards on the race course. Race administrators seek to shield themselves from liability for such injuries by requiring competitors to sign release forms.

The purpose of this evaluation is to define the impact that the release or waiver has on bicycle and other accident reconstructions.

Waiver Language and Those Areas of Accident Reconstruction Covered by the Waiver

A release usually contains language similar to the following:

> I hereby waive, release and discharge any and all claims and/or damages for death, personal injury or property damage which I may have, or which may hereafter accrue to me, as a result of my participation in said event. This release is intended to discharge the sponsoring entity from and against any and all liability arising out of or connected in any way with my participation in said event, even though that liability may arise out of negligence or carelessness on the part of the persons or entities mentioned above.

By signing this type of release, the participant agrees to "hold harmless" the race organizers for any injuries sustained during the event, even though the organizers may have been negligent in managing the race. Such releases have, however been successfully challenged when an accident reconstruction has clearly shown fault on the part of the race management. The attorney or engineer must be aware when reconstructing the accident that violation of bicycle racing safety standards may make the race organizers liable, despite a signed release. Two landmark decisions from the appellate courts in California and Virginia illustrate the boundaries of a release when industry standards are violated during race administration.

Albert I. Bennett, Plaintiff and Appellant
v.
United States Cycling Federation et. al.
Defendants and Respondents,
193 Cal. App. 3d 1485, Cir. R022865,
Court of Appeal, Second District, Division 5,
August 4, 1987.

The amateur bicyclist, Bennett, was severely injured when the race organizers allowed a car onto the closed race course. The race course was completely sealed off by barricades. In addition, race participants were informed by literature and instruction that the course was closed and no motor vehicle traffic would be permitted on the course. However, a course marshall granted a motorist's request for entry onto the course during the race. After the marshall removed a barricade and the motorist entered the course, the peleton (large group of riders) was unable to avoid crashing into the motor vehicle due to the normal operation of racing. Bennett was severely injured in the crash. He subsequently sued the race organizer for negligence in allowing the motor vehicle onto the course.

Bennett's counsel argued that the race organizer's actions were not covered by the waiver, because the accident was an unexpected risk, not the type of risk contemplated by the releasing party. The court agreed, finding that the waiver only released the organizers from hazards that were obvious to or reasonably foreseen by the race participant. Since the course was known to be closed to automobiles before the race, it is doubtful whether Bennett, or any participant, would have realistically appreciated the risk of a car traveling in any direction along the race course. Consequently, the risk was not covered by the waiver.

Robert David Hiett
vs.
Lake Barcroft Community Association, et. al.
Record No. 911395,
Opinion by Justice Barbara Milano Keenan,
June 5, 1992,
Circuit Court of Fairfax County, Fairfax, Virginia.

At the beginning of the Lake Barcroft Teflon-Man Triathlon, Robert Hiett was rendered a quadriplegic when he dove into the water at the start of the swimming event. Hiett waded into the water to a point where the water reached his thighs. When the race started, he dove into the water and struck his head on the lake bottom or an object beneath the water surface.

Hiett alleged that the organizers failed to ensure that the lake was reasonably safe, properly supervise the swimming event, advise the participants of the risk of injury, and train them how to avoid such injuries. The court agreed. Citing Johnson's Admix v. Richmond and Danville R.R. Co. (1980) (an agreement entered into prior to any injury, releasing a party from liability for negligence resulting in personal injury, is void because it violates public policy), the court held that the release violated public policy.

The Importance of a Waiver in Reconstructing Bicycle Racing Accidents

The above case law indicates that a race organizer may be liable for injuries sustained during a race, despite a signed waiver, if the organizer violated bicycle racing standards. These standards are defined by the USCF and the Triathlon Federation ("TriFed"). A review of these regulations reveals the definitions discussed below.

Reasonably Foreseen Hazards Covered by the Waiver

In bicycle racing, reasonably foreseen hazards include: collisions with other riders, negligently maintained equipment, bicycles that were unfit for racing but were nevertheless passed by the organizers, bad road surfaces, and traffic that may be present on an open course. In swimming, these hazards include: collisions with other swimmers, adverse water conditions normally encountered by swimmers, and physiological failure on the part of the swimmer. In running, these hazards include: collisions with other runners, bad road surfaces and traffic that may be present on an open course.

Unforeseeable Hazards Not Covered by the Waiver

In bicycle racing, unforeseen and unreasonable hazards include: negligently maintained race courses, improper traffic monitoring on race courses, and improperly maintained crowd control when crowd control is guaranteed to the contestants. In swimming, these hazards include: failure to insure a safe swimming course (including the failure to remove underwater hazards). In running, these hazards include: failure to maintain traffic control when traffic control is guaranteed to the contestants.

Conclusion

The recent court decisions discussed here have redefined the scope of an accident reconstruction involving a signed release or waiver. A waiver no longer necessarily precludes a finding of liability on the part of the race organizer. Thus, the attorney or engineer must now determine whether a causal factor of the accident was a violation of bicycle racing safety standards (as defined by the USCF and TriFed). If this question is answered affirmatively, the waiver will likely not shield the organizer from liability.

Chapter 28

An Analysis of the Engineering Dynamics of a Bicycle Falling Sideways

Introduction

A cyclist's loss of balance is familiar to anyone who has fallen off a bike. When balance is lost, the cyclist falls to the side. This phenomena has never been the subject of empirical analysis in the engineering literature due to the seemingly obvious simplicity of the problem. However, defining the phenomena as a function of the engineering dynamics involved is necessary for reconstructing bicycle accidents.

The purpose of this analysis is to define two methods for illustrating how a cyclist loses balance. A cyclist's loss of balance is frequently a contributing factor in cycling accidents. Thus, these methods can be used to determine whether the cyclist's loss of balance caused the accident, rather than an alleged defect in the subject bike and/or roadway.

Engineering Analysis

It is easier to balance on a bicycle that is moving than on a bicycle at rest. Rotating wheels have angular momentum. Thus, flipping rotating wheels means a change in the angular momentum, Thus, requires a greater torque than tipping wheels at rest. However, by removing the wheel rotation issue from the analysis, the engineering dynamics can focus on the cyclist's balance. The basic assumptions employed in the analysis are:
1. the bike tires do not slip on the pavement,
2. the bike is not steered, and
3. there are no gyroscopic forces on the cycle.

Sum of Forces, Method 1

If the forces distributed around a bicycle are as represented in Figure 1, then a trigonometric relationship can be formulated. The distribution of the forces is based on the assumption that the cyclist is seated. The forces are:

F_{RH} = force placed on the right handlebar by the right hand
F_{RF} = force placed on the right pedal by the right foot
F_S = force placed on the seat by the rider in the seated position
F_{LH} = force placed on the left handlebar by the left hand
F_{LF} = force placed on the left pedal by the left foot
N = the opposite and equal force from the surface or mass of bike plus the mass of the rider

If the cycle is in equilibrium, or properly balanced, then;

$$\Sigma F = F_{RH} + F_S + F_{LH} + F_{RF} + F_{LF} - N \text{ and } \Sigma F = 0$$

To describe a cycle tipping to the right as viewed from the front;

$$+ \Sigma \tau = I \alpha$$
$$= F_{LH} R_{LH} \cos \delta_{LH} F_{LF} R_{LF} \cos \delta_{LF} + F_S \sin \theta - F_{RH} \cos \delta_{RH} - F_{RF} \cos \delta_{RF}$$

To describe tipping in either direction, the summation of Tau (τ) can be used as follows:

If

$\Sigma \tau > 0$ the bike tips to the left
$\Sigma \tau < 0$ the bike tips to the right
$\Sigma \tau = 0$ the bike is stable

Analysis of Center of Gravity, Method 2

An example of this method is shown in Figure 2. This method focuses on the cyclist's center of gravity, C_g, which is typically located at approximately the cyclist's navel. Therefore, the relationship between C_g and mass is:

$\Sigma F = 0$

$\Sigma F = [M_{bike} + M_{rider}]g - N$

where $N = M_{bike} + M_{rider}$ in the opposite direction

If rotation occurs, then;

$+ \tau = I \alpha$

$= [M_B + M_R]g$ = vector in a plumb line after the movement to the right is initiated[1]

Similar to Method 1, the following relationships are developed;
If

$\Sigma \tau > 0$ the bike tips to its left

$\Sigma \tau < 0$ the bike tips to its right

$\Sigma \tau = 0$ the bike is stable

Summary of Tipping Analysis

A cyclist's loss of balance may seem obvious to the casual observer. However, a quantitative explanation of the phenomena is needed for reconstructing bicycle accidents. The methods defined above can provide the necessary quantitative explanation. A forensic engineer who wishes to employ the methods should note the following salient points drawn from the tipping analysis:

1. The centrifugal forces imparted by the rotation of the wheels will make a bicycle more stable. As a result, the amount of displaced force in both Method 1 and Method 2 must, by necessity, be greater when the wheels are rotating. Field studies reveal that approximately four times the amount of displaced force is needed for loss of balance during wheel rotation at 10 mph.[2] As expected, a moving cycle has a higher stability level than one that is at rest.

2. Observation of a cyclist pedaling a bike reveals that the cyclist's weight is transferred from one side of the bike to the other as a function of the downstroke. This phenomena is even more apparent when a cyclist is observed climbing a steep grade. Loss of balance occurs when the cyclist's weight is transferred to the side beyond the capacity of the cyclist to compensate to the other side. In the field, this happens when lateral forces are transmitted to the cycle.

Conclusion

The above principles and methods may be used to determine the causal factor of a sideways fall in a bicycle accident reconstruction. In cases involving such falls, it is often alleged that a rough area on the road or bike path caused the cyclist to fall to the side. However, my experience in reconstructing these types of accidents has shown that such allegations often cannot be substantiated. A rough area, or a sudden steep incline, will transmit a radial force to the cycle.

But a lateral force, rather than a radial force, is necessary to unbalance the cycle.

Endnotes

1. It should be noted that this is at a given time (T) in the rotation. Differentiation of the equations over time or d /dT is needed to fully show rotation at all times.
2. Force on a pedal in the down position was measured on a simulator consisting of a pair of rollers. Displaced force was utilized by determining a weight differential. The factor of four is considered to be a conceptual value. A great deal of additional work needs to be done in this area before any absolutes can be given. However, this initial evaluation indicates that the relationship between displaced force, loss of balance and velocity is curvilinear.

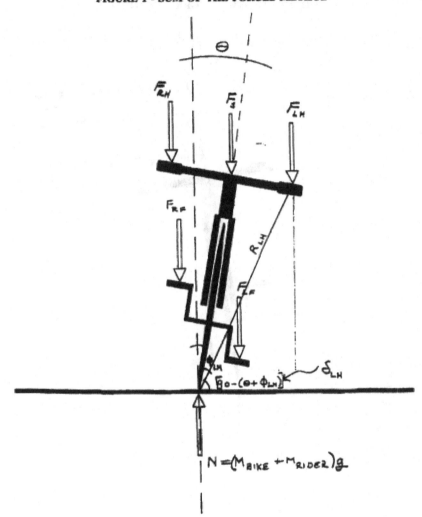

FIGURE 1 - SUM OF THE FORCES METHOD

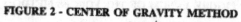
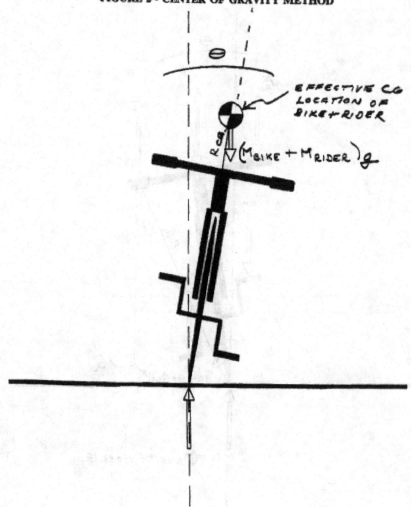

FIGURE 2 - CENTER OF GRAVITY METHOD

Chapter 29

Metallurgy and the Reconstruction of Bicycle Accidents

Introduction

The purpose of this chapter is not to describe the field of metallurgy in detail. There are many fine Engineering text books that can be utilized to educate the forensic Engineer in this area. It is assumed that the Engineer has a working knowledge of metallurgy prior to the use of the techniques described in this chapter. I have presented areas of bicycle accident reconstruction where the field of metallurgy is applicable for determining the causal factors of bicycle accidents. The major areas of emphasis will be: frame failures, component failures, angle of impact between a motor vehicle and a bicycle, and handling characteristics that are potentially germaine to cycling accidents.

Frame Failures

Titanium

The bicycle industry has undergone a dramatic revolution in frame materials in the five years leading up to the publication of this book. One of the most exotic and popular metals among cyclists is the use of titanium alloys. Prior to the advent of companies such as Litespeed, the common bicycle frame material was made from steel alloys. The density of a steel alloy such as Reynolds 753 is 0.285 lb/in^3 with an elastic moduli of 30 million psi. With the advent of titanium alloys, the density decreases to 0.165 lb/in^3 while the elastic moduli only decreases to 17 million psi.

As can be seen from this data, both titaniums' density and modulus of elasticity is approximately halfway of that of steel. Titanium, when properly treated, gives a stiffness that is close to steel, but a weight that is approximately 32 % lighter.

Litespeed Titanium Quality

In investigating the causal factors of frame failures, knowledge of the strength versus weight of the frame is critical. The best made high end price frames in the bike industry are made 6AL-4V CW Titanium. A comparison of this type of Titanium has been made by Litespeed[1]. In their analysis of the various materials for frame manufacture, they considered yield strength divided by density. The higher this value, the stronger the material per density unit.

Table 1
A Comparison of Strength Versus Weight of Materials Used in Bicycle Frames

Material	Yield Strength Divided by Density
6AL CW Titanium	884,000
3AL-2.5V CWSR Titanium	710,000
Scandium	564,000
7005 Aluminum	455,000
Pure Titanium	432,000

As can be seen from the data, the 6AL CW Titanium used in Litespeed bicycles gives an extremely strong frame with a light weight not seen in other frame materials.

Aluminum Frames

These frames are typically made by companies like Klein or Cannondale. Pure aluminum is never used for bicycles frames. The frames that are marketed as aluminum are actually aluminum alloys. There are many different codes in defining the type of aluminum in a bike frame. Basically the industry uses a code system that differentiates between wrought alluminum alloys and caste aluminum alloys.

Wrought aluminum alloys use a code usually ending in a series from 1XXX to 7XXX. The industry typically uses the following code designations to build bicycles:
- Code 2014 contains 4.5 % Copper
- Code 6-61 contains 1 % Magnesium and 0.6 % Silicon
- Code 7075 contains 5.6 % Zinc.

The actual composition and treating characteristics of each manufacturer's process is a propreitary procedure. Most manufacturers utilize heat treating to

achieve higher strength. While the procedures can be proprietary, the standard general designation of heating, or temper, is:
- O for annealed
- T4 for heat treated and aged at room temperature
- T6 for heat treated and aged at elevated temperature.

The yield strength for the highest strength temper of aluminum alloys is approximately 73,000 psi - 78,000 psi. As noted in the previous section, the light weight of alluminum does make it an excellent source for frames, but it is somewhat limited as compared to Titanium or Steel by a lower yield strength. In addition, the quality control in the manufacturing of aluminum is extremely important due to the heat treatment procedure. A new Columbus 6000 alloy Aluminum tubing has come on the market which increase the lower yield strength of aluminum. The addition of this new tubing is noted in the next section.

Steel Frames

As most Engineers are aware, steel is used in most structural applications. The bike industry utilizes only a few types of steel in the manufacture of bicycles. From the manufacturers literature[2], the various yield strengths' can be compared to the more common tubing used in the manufacture of bicycle frames.

Table 2
Yield and Tensile Strength for Steel Frame Tubing

Generic Brand	Yield Strength lb/in^2	Tensile Strength lb/in^2
Columbus 6000 alloy	(a)	(a)
Excell	171,000	200,100
Super Vitas 983	128,000	156,500
Reynolds 753	156,800	179,200
Reynolds 531	100,800	116,500
Columbus	107,000	121,000

(a) This alloy is included in the list since it is new (2001) on the tubing market. This alluminum alloy is advertized as having the advantages of the steel alloys but not the disadvantage of the lower 73,00-78,00 lb/in^2 yield strength of the other alloys. (See Bicycle Retailer & Industry News, March 15, 2001, pg. 20).

Exotic Materials

Due to the growth in triathlon competitions, the bike industry has come forward with several exotic composite materials. Most of these materials are continually evolving and what is considered state of the art in the year 2001 will evolve into something very different in five or more years. Some of the more exotic frame materials are given here so the Forensic Engineer will recognize the material in the appropriate setting.

Baron Carbide

This material originally surfaced in the United States Army as material on Apache helicopters as a light weight replacement for aluminum bullet proofing. Griffen USA currently uses the baron carbide and manufactures an advanced ceramic metal matrix composite by pressing the powder together. The Griffen triathlon bike is used by several top racers. The baron carbide matrix is substantially more expensive than titanium or aluminum.

Yield and tensile strengths are currently unavailable on this material. The modulus of elasticity is 14.3 million psi which is approxiamtely 4.2 million psi greater than aluminum.

Advanced Composites

The first ever finite element analysis of the structural forces at work on a bicycle frame were conducted in 1986 by Kestrel. This company then applied the analysis to the use of proprietary components in the manufacture of bicycle frames and forks. One of the advantages of advanced components is that the frame can be made in one piece. As a result, there are no joints in the finished frame. This results in an incredibly strong frame with no weak areas that could be subject to failure. It is very doubtful, from a Forensic Engineering prospective, that frame failures on one piece composite frames, such as Kestrel, are the result of design or manufactuing error.

Forensic Engineering Analysis of Frame and Fork Failure

From the continuing advance of the Bike Industry in finate frame analysis as well as composite material use, the analysis of the causal factors involved in failure have evolved substantially in the last five years. When presented with a frame failure, the Forensic Engineer should concentrate the analysis in the following areas:

1. <u>Frontal Forces</u> - Elsewhere in this book, the sequence of frame collapse during the application of static and dynamic forces are noted. Due to the high cost of most composite frames, individual frontal force testing of these frames is prohibitive. However, from the modulus of elasticity values and yield strengths of these various materials, a design defect can usually be ruled out by the Forensic Engineer. If a frame failure with the type of bonding noted in Chapter 24 occurs, the engineering analysis should center on some type of outside factor causing failure.

2. <u>Fork Deflection</u> - Previous work in this text plots velocity versus fork deflection (See a Determination of Bicycle Speed by using Crush Data). This profile appears to be still accurate as of 2001 for most low-end steel frames. With the advent of advanced composite frames and forks, each frame deformation must be handled on an individual basis. Most accidents that the author has reconstructed in the past five years yields frame deformation different from the the work accomplished in the speed versus crush data section. For example, the earlier forks on road bike frames tended to be extremely durable during a crash. It would not be unusual to see the top tube and down tube broken or crimped and the front fork not showing any appreciable damage. Since the advent of composite forks, the forks tend to collapse prior to the top tube or downtube in front collisions. This in no way suggests that composite forks are not safe or adequate in normal use. It does appear that composite materials do not handle severe decelerated forces as well as the older steel forks. The Forensic Engineer should attempt to duplicate the frontal impact with an exemplar frame in the laboratory. Failing that, speed of impact would best be determined from calculating body trajectories as is defined elsewhere in the text.

3. <u>Speed of Impact</u> - The Forensic Engineer can safely apply the data points developed elsewhere in this book in determining speed of impact and angle of impact of a cycle with a motor vehicle. With the advent of composite materials, independent crush profiles need to be determined in the laboratory for the bicycles. However, since many of these frames can cost $2000 or more, laboratory testing is very cost prohibitive. A better method of analysis for developing impact data, is the use of trajectories of the accident victim as well as point of rest. That information has been developed elsewhere in

this book and has been shown to be very accurate by the engineering commmunity in the last five years of reconstructing collisions between motor vehicles and pedestrians, bicycles and motorcylists.

Metallurgy and Component Failure

Having investigated thousands of bicycle accidents, the author has found the causal factor of these accidents to usually be something other than component failure. However, component failures do occur and should be analyzed as a potential cause of the accident.

Front Fork Failure

The predominant failure mode of these accidents is the cracking or breaking of the fork crown or related components. The usual failure occurs on an off road cycle such as a mountain bike or BMX bike. These cycles are subjected to a variety of deccelarated forces from jumping and other off road uses.

In the time period since 1995, until the present, several front fork recalls have occurred in the cycling industry. Most of these recalls dealt with improper welding at the joint of the fork tube and the crown, collapse of the front fork blades at the juncture with the fork crown, and breakage of the fork blades due to improper wall thickness. In every failure that this author personally investigated, the causal factor was the improper welding or manufacture of the component offshore in Taiwan or China. In all instances, the United States distributors or manufacturers properly designed the subject components. The properly provided specifications and designs were found to be inadequately followed by the offshore manufacturers.

Conclusion

The key to determining the causal factor, in bicycle component failure where metallurgy is the potential issue, is adequately analyzing the original design drawings and specifications. Once this analysis is complete, a comparison should be made between the actual design and the component at issue. In almost all of the failures that this author has investigated, it has been the failure of the component or frame manufacturer to follow the proper provided design and specifications.

Endnotes

1. http://www.litespeed.com/html/technology/titanium.html. This web site was effective in June 2001.
2. All of these manufacturers maintain web sites.

Chapter 30

Determining Total Reaction Time for Accident Reconstruction and Civil Engineering Design

Introduction

In reconstructing accidents, a great deal of confusion exists as to the proper application of total reaction time to the entities involved in the accident. The purpose of this paper is to review the existing literature and to connect the various professions' definition of total reaction time. The connection being made in this paper is to establish total reaction time (TRT) that utilizes the principles of conspicuity, as well as civil and traffic engineering. Confusion is very apparent in the reconstruction of accidents and in engineering design over the use of conspicuity. The ability to see a stimulus, under ambient conditions present, is the study of conspicuity. The use of factors such as color, lighting, reflectivity and other environmental criteria must be considered to determine reaction time.

Basic Human Perceive React Time

In Paul Olson's excellent paper[5], the components of perceived reaction time (PRT) are defined. Detection begins when some object or condition of concern enters the driver's field of vision. This detection step concludes when the driver develops a conscious awareness that "something" is present. Olson goes on to conclude that this stimulus may be in the field of view of the driver for some time before it is detected. The driver then moves on to identify the object, decides the appropriate action and commands the muscle groups to carry out the required action. Olson's analysis continues in an attempt to summarize the research so that the time required for a driver to respond to an emergency can be quantified. The analysis correctly confirms that this is very difficult to accomplish. The reason is that, among other things, there is a great deal of variability in the conspicuity of the stimulus. Olson goes on to compare many studies on this issue and concludes that a basic response time of 1.5 seconds would be appropriate. Olson does state that an increase in the PRT of 1.5 seconds is, unfortunately, a matter of judgment [6,7,8].

Examination of the data in the various studies presented in Olson's paper indicate that the actual human physiological response time is between .75 and 1.5 seconds dependent on age and the stimulus used in the study. The problem in detailing a finite response time is the other variables that are present in the real world. Foremost of these variables presented in the studies [6,7,8] supporting the Olson paper is the ability for the stimulus to be seen. This issue of conspicuity is interwoven throughout the studies referenced in the Olson paper. The PRT values given by the different studies also are developed during ambient daylight hours.

Perceive React and Avoid Time at Night

Olson and Sivak[4] broaden the problem of perception to include nighttime driving. Having reconstructed hundreds of pedestrian and cycling accidents at night, I can personally verify that the variables are fairly complex and are a function of the clothing worn by the pedestrian or cyclist, as well as other factors. Since the usual accidents investigated include a motor vehicle colliding with a pedestrian or cyclist, the emphasis in Olson and Sivak's analysis is the conspicuity of the individual within the range of headlights. Since the speed of approaching motor vehicles may vary, Olson and Sivak give response time in response distance to pedestrian targets. They include the far range of the spectrum and use dark and light targets. The expected response distances in Olson and Sivak's paper are as follows:

Table 1: Expected Response Distances (FT) to Pedestrian Targets

Headlamp Beam	Target Location	Clothing	Average Response Distance	Range
Low	Right	Dark	80	0-160
		Light	160	80-240
	Left	Dark	60	0-120
		Light	120	60-200
High	Right	Dark	120	60-300
		Light	240	160-400
	Left	Dark	120	60-300
		Light	240	160-400

Olson and Sivak also note that the use of high beams at night does help but not as much as "one might think". Their rough rule of thumb gives high beams improving detection – identification distance to objects on the right side of the road by 50 percent and to objects on the left side of the road by about 100 percent. This difference is due to the fact that low beams direct most of their intensity to the right. Olson and Sivak also note that proper reflective clothing will increase the response distance substantially, but do not quantify the difference.

In order to determine a total reaction time or TRT, the review of the literature reveals that the stimulus must be recognized as actually being present. Olson's use of the 1.5 seconds as a basic response time is a solid first step. As civil and transportation engineers [1] realize when designing intersections, a line of sight reaction time of 2.5 seconds for the design speed is typically used for the minimum time needed.

Regardless of the reaction time used, it is obvious that the stimulus must be recognized as being something requiring a driver response.

The Blomberg Studies [2,3] consists of the best methodology that has been employed in field studies to define the effect color and retroreflectivity will have on the initial recognition distance. During nighttime hours, Blomberg found that the use of lighting, reflective clothing and bike retroreflectors dramatically increased the recognition distance from as low as 75 feet to as much as 1200 feet. Independent investigators have done verification of Blomberg's work. His analysis has been found to be correct and the increased recognition distances have been found to be reliable [4].

The salient points in Blomberg's studies that have a direct bearing on the development of a TRT are as follows:

- Nighttime recognition distances for an object in dark clothing with no other means of conspicuity are approximately 75 feet on low beams. High beams do not appreciably increase recognition distance (See Table 1).
- Nighttime visibility is increased to as much as 560 feet with fluorescent clothing. [2, 3, 4] The visibility is further increased by lighting systems that meet the DOT requirement of 600 feet. Beyond the 600 feet of car beam headlights for both high and low beams, recognition distance of an object should be determined by using unbiased observers as was done in the Blomberg studies.
- Daytime visibility is increased from 400 feet to 2200 feet with fluorescent clothing [2,3,4]. The other types of clothing evaluated in the Blomberg studies, as well as other researchers [2, 3, 4], clearly show visibility and hence recognition distances will vary tremendously with the type of clothing worn.

Conclusion

The representative literature discussed in this paper reveals that certain points are universal among the Psychology, Optometry and Engineering Professions regarding TRT. These points are as follows:

1. The human body basically reacts physiologically from .75 to 1.5 seconds in expected stimulus response situations. An example of this scenario is the measuring of response time for someone to react to a sudden flash of light in the laboratory.
2. The human body takes varying periods of time to recognize that a response to a stimulus may be necessary. An example of this scenario would be a driver in a motor vehicle attempting to discern and identify an object in the road ahead at night. For

example, assume a driver comes upon a cyclist in the road ahead at night. The cyclist is dressed in black and has no lights or retroreflectors on the cycle. In this instance, the driver will not discern movement sufficient enough to alert him or her to the need to avoid until close to 75 feet from the eventual point of impact.

3. A definition of TRT is the total time a subject will take to perceive, react and avoid a stimulus. When the Forensic Engineer is determining the causal factor of an accident, the perceive time must take into account the environmental and ambient conditions that are present. This can only be done from a properly calibrated accident site. The calibrated site must utilize the accepted methods for accurately defining, with a high degree of engineering and scientific certainty, the conditions at the scene when the accident occurred.

4. The Engineer should then attempt to reproduce the conditions utilizing unbiased subjects to accurately define the perceive distance. If an accident occurred at night on low beams with an accident victim dressed in a certain manner then that is the situation that should be reproduced using unbiased subjects. From that simulation, the perceive time can be determined.

5. Once an actual perceive time is determined, the TRT can be defined. If for example, under the simulation, it is found that perception time is 3 seconds at a given speed, the TRT would be equal to:

6. Perception time + reaction time + avoid time or 3.0 seconds + 1.5 seconds + 1.0 seconds = 5.5 seconds.

Summary

The review of representative literature reveals that perception is a function of the conspicuity of the subject under the ambient environmental conditions in place at the time of an accident. The definition of the proper perception time should be determined from simulation of the actual environmental conditions and the conspicuity of the accident subject at the time of the accident. This should occur using the actual closing speeds of the individual components of the accident. If a motor vehicle is approaching a subject at 30 MPH, then that is the speed that should be used to determine perceive time [a]. If the actual total reaction time (TRT) is used for each accident scenario being investigated, the causal factors of the accident can be more accurately determined.

[a] I actually witnessed a "accident reconstructionist," simulate on video a cyclist leaving the pool of low beam light in front of a truck at night. This simulation was alleged to represent the lighting conditions while the truck was approaching the cyclist at 35 MPH. The " accident reconstructionist," then swore under oath that the 22 seconds it took for the cyclist to disappear from this pool of low beam light actually represented the lighting conditions, and the ability of the cyclist to be seen, at the time of the accident.

Endnotes

1. AASHTO. A Policy on Geometric Design of Highways and Streets. *American Association of State Highway and Transportation Officials*, 1990, pp. 45-49.
2. Blomberg, R. D., A. Hale, and D. F, Preusser. Conspicuity for Pedestrians and Bicyclists: Definition of the Problem, Development and Test of Countermeasures. *U.S. Dept. of Transportation, National Highway Traffic Safety Administration*, DOT HS-806-563, April 1984, pp. i-xii, 1-78, Appendix A-1-K-5.
3. Blomberg, R. D., A. Hale, and D. F. Preusser. Experimental Evaluation of Alternative Conspicuity – Enhancement Techniques for Pedestrians and Bicyclists. *Journal of Safety Research*, 1986, Vol. 17, pp. 1-12.
4. Green, J. M. *Bicycle Accident Reconstruction*. Iuniverse.com Press, 2000, pp. 59-70, 77-82, and 85-87.
5. Olson, P. L., PhD. and M. Sivak, PhD. Visibility Problems in Nighttime Driving. *Human Factors and Biodynamics*, 1981, Chapter 15, pp. 383-401.
6. Olson, P. L. *Driver Perception Response Time*. The University of Michigan Transportation Research Institute, SAE paper No. 890731, 1989.
7. Owens, A. D., E. L. Francis, and H. W. Lebowitz. *Visibility Distance with Headlights: A Functional Approach*. SAE paper No. 890684, 1989.
8. Sens, M. J., P. H. Cheng, J. F. Wiechel, and D. A. Guenther. *Perception/Reaction Time Values for Accident Reconstruction*. SAE paper No. 890732, 1989.

Chapter 31

The Causal Factor of Bus Wheel Injuries And a Remedial Method for Prevention Of These Accidents

Introduction

The accident statistics for injuries caused by pedestrians or cyclists being injured, or killed, by U.S. transit buses have typically been categorized simply as either fatalities or serious injuries[1]. Although anecdotal information from police accident investigators and Forensic Engineers have indicated that certain types of accidents with transit buses are more prevalent than other types, definitive data has been lacking. Recent risk management efforts at various transit authorities[2] have revealed a prevalent type of accident from transit vehicles interacting with either cyclists or pedestrians. The predominant accident type seems to be pedestrians or cyclists being pulled into the bus-wheel, as opposed to individuals being struck by the vehicle body[3]. Further questioning of transit personnel indicates that, in most cases, the accidents occur from the rotating bus transit wheel on the bus as it passes the individual as opposed to the cyclist or pedestrian running into the stationary transit vehicle or tire. Surprisingly, the type of accident where the bus strikes the cyclist or pedestrian in an area other than on the rotating wheel is almost negligible.

While statistical reporting and analyses of this data has not been accomplished to a high degree of engineering certainty, most risk managers for metropolitan transit authorities will admit to a surprisingly high number of these rotating wheel type of accidents[4]. By whatever analysis method that is used, there is a clear problem with these types of accidents. Of particular interest is the fact that most points of impact onto the bus body appear to occur at the point of the rotating wheel in the bus wheel well.

The analysis in this paper is focusing on Transit Authority Buses since risk assessment managers have identified high incidents of injury at the site of the rotating wheel for these vehicles. More probably than not, other types of motor vehicles, such as trucks, would also tend to have a high degree of cyclist or pedestrian accident prevalence at wheel wells. Currently the Engineering, and related literature, does not contain valid statistics on wheel well accidents other than Transit Authority vehicles. As a result, this analysis centers on these vehicles, but can be applied to other heavy-duty vehicles as well.

Bernoulli's Principle

Since transit authority personnel agree that there is a problem with pedestrians and cyclists being impacted in the proximate vicinity of the wheel well, an explanation is needed for this set of data. If Bernoulli's Principle is defined in terms of pressure, the equation becomes[5]:

$$\frac{1}{2}(Rho)V^2 + p + (Rho)gy = constant$$

where: Rho = density
p = pressure
g = acceleration due to gravity
y = elevation.
V = velocity

If y does not change, then an increase in V means a decrease in p. This basically means that as a transit authority bus passes a cyclist or pedestrian at a higher speed, there will be a decrease in pressure between the two entities. Since the bus is at a much higher mass, the pedestrian or cyclist will be drawn toward the vehicle.

This does not explain why most points of impact occur on or near the rotating wheel. If we assume that elevation (y) and the constant (k) are both 1 and the equation is unitless[6], then the relationship between pressure and velocity becomes:

Solving For P

$$P = \frac{1}{\frac{3}{2}V^2 + 33.3}$$

The plot of P^{-1} versus V is shown in Figure 1. If a bus passes a pedestrian or cyclist at 10 MPH or 14.7 ft/second, the rotating wheel of an assumed 6 ft of circumference will rotate approximately 2.5 revolutions/second[7]. Therefore, regardless of the units used, if you compare the different speeds of the rotating wheel at 10 MPH, there is a ratio of 2.5/1 of wheel velocity to bus speed. For comparison purposes, the inverse pressure (p^{-1}) from Figure 1 is 183 at 10 MPH while the inverse pressure for the increased speed of the bus wheel is 971. From this unitless analysis, it is obvious that regardless of the method used, the rotating wheel of the bus, or any large vehicle, will create a low pressure between the cyclist or the pedestrian that is vastly different than just the motor vehicle passing the individual. As a result, there is a greater potential for the cyclist or pedestrian to be pulled into the motor vehicle body. This lower pressure resulting from the higher rotational velocity of the motor vehicle wheel explains the greater grouping of the points of impact at the wheel well of transit authority buses.

Figure 1. Velocity (mph) v. Inverse Pressure (psi)

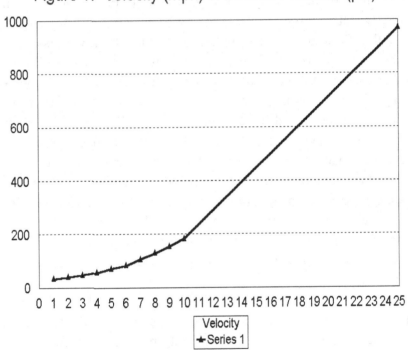

The constant K and elevation H are assumed to be 1.

An increase in velocity means a decrease in pressure and accounts for the fact that passing ships run the risk of a sideways collision. This is due to the fact that water flowing between ships travels faster than water flowing past the outer sides. Therefore, water pressure acting against the hulls is reduced between the ships. Unless the ships are steered to compensate for this, the greater pressure against the outer sides of the ships forces them together.

In order to design a remedial measure to prevent the inordinate amount of accidents at transit bus wheels, Bernoulli's principle must be utilized.

A Practical Design for Preventing Wheel Well Accidents

There are two problem areas in designing a remedial measure to prevent wheel well accidents.

The first area of concern is the description of the low-pressure gradient between the rotating high velocity bus wheel and the pedestrian or cyclist. The second area of design application is the prevention of the physical entrapment of the cyclist or pedestrian from a bus turning into the path of travel of either entity. In this second area, physical entrapment can also occur from the low-pressure gradient pulling the cyclist or pedestrian to the physical proximity of the rotating wheel.

Bernoulli's Principle can be applied to disrupt the low-pressure gradient that can pull a cyclist or a pedestrian into the high velocity-rotating wheel by considering the lifting force of the airplane wing. The airfoil of a curved airplane wing adds considerably to lift and results in a greater difference in pressure between the lower and upper wing surfaces. This net upward pressure multiplied by the surface area of the wing gives the net lifting force. By having a curved wheel guard at the forward leading edge of the transit bus wheel well, a net outward pressure away from the direction of travel of the bus is produced. This results in the complete elimination of the low-pressure gradient that would draw the cyclist or pedestrian into the high velocity-rotating wheel. More importantly, the curvature of the guard would act like an airplane wing and literally be able to push the cyclist or pedestrian out of the path of travel of the transit bus.

As noted in Figure 2, a wheel well guard with a leading edge capable of lifting the air outward from the bus's path of travel is shown. By utilizing this curvature, the cyclist or pedestrian is actually pushed away from the leading edge of the wheel well by the outward change in air gradient. The strength of materials of the guard should also be capable of actually pushing a pedestrian or cyclist away from the path of travel of the transit bus if the individual falls into the path of the rotating wheel.

Figure 2. Bus Wheel Well With S-1Gard

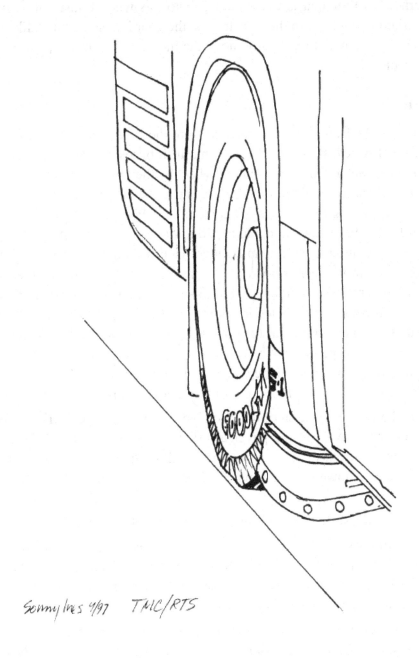

As noted in Figure 2, field trials do support the ability of this design to physically move a subject from the path of travel. This is helpful in instances where Bernoulli's Principal is not a causal factor, as when transit buses turn into pedestrians or cyclists. In those instances, the guard must act much like the cowcatcher on a train and physically move the individual from the path of the rotating wheel.

Conclusion

As described in the Bernoulli analysis, and from the field data, the causal factor of most cyclist-pedestrian accidents with transit buses are from the individuals either being dragged into the rotating wheel by the lower pressure gradient or from the physical impacting of the bus during a turning radius.

When investigating these types of accidents, the Forensic Engineer should realize that Bernoulli's Principal could be a definite causal factor. Also, the bus physically turning into the path of an accident victim should be considered. Of equal importance in the analysis is the fact that remedial measures are easily available to prevent these accidents. The illustrated S-1 Gard (generic name), see Figure 2, has been implemented in several municipality's. Thus far, in those municipalities that have initiated this program the accident rate has decreased from several incidents per year to zero.

Case studies are being developed at: Washington D.C., Los Angeles, California, Miami-Dade County, Florida, and San Diego, California. Ferrone has accomplished a field evaluation of the effectiveness of the S-1 GARD[8]. In that effort, the emphasis was on the physical effectiveness of the S-1 GARD in moving a stunt man out of the path of the rotating wheel. The experimental runs, as expected, showed that the physical properties of the S-1 GARD did successfully remove the individual from the path of the rotating wheel.[9]

Additional case studies are being planned. The effectiveness of the S-1 GARD [10] to eliminate the low-pressure gradient at wheel wells as a function of speed is needed. Theoretically, the effectiveness of the S-1 GARD should increase as velocity increases. An additional study to determine the effectiveness of the S-1 GARD on heavy-duty vehicles should also be considered.[11]

The use of a guard on the rear wheel wells of transit authority buses as well as heavy-duty vehicles is in its infancy. The initial evaluations clearly show a dangerous problem exists. Field studies conducted thus far yield excellent preliminary results in utilizing the S-1 GARD, the only guard currently on the market, to completely eliminate wheel well accidents. Hopefully, the release of this paper into the Forensic Engineering community will enable the reason behind these accidents to be acknowledged as well as the remedial measure needed to eliminate the problem.

Endnotes

1. US Department of Transportation, National Highway Traffic Safety Administration, "Traffic SafetyFacts," for Pedacyclists (DOT HS 808 957) and Pedestrians (DOT HS 808 95).
2. See Bus Wheel Injury Study, San Diego Transit Corporation, DART, G. Transit Richmond 1989-1993.
3. Ibid Although Risk Managers at these transit authorities have not applied statistical analyses, the predominant, and in most cases, total accident rate is by the rotating transit vehicle wheel.
4. Most reconstruction projects that I have had with accidents involving buses, pedestrians and cyclists did, in fact, involve rotating bus wheels. In my own interviewing of transit personnel, this type of accident is definitely the most common.
5. Hewitt, Paul G. "Conceptual Physics, Seventh Edition," City College of San Francisco, Herpa Collins College Publishers, pp. 239-242.
   Serway, Raymond & Bechner, Robert J., "Physics for Scientists and Engineers, Fifth Edition," Saunders College Publishing, pp. 469-476.
6. We are interested in making a comparison of the increased velocity of the bus wheel versus the body of the bus passing a cyclist or pedestrian. The important issue here is the inverse relationship between pressure and velocity.
7. $14.7/6 = 2.45$ rps.
8. Ferrone, Christopher W. "A Field Evaluation of the S-1 Gard: Transit and Shuttle Bus Applications", Society of Automotive Engineers, Inc., Paper No. 982775, 1998.
9. Public Transportation Safety International Corporation, US Patent # 5,462,324 WO 95/28300, S-1 GARD, Pacific Center, 523 West 6th Street, Suite 1222, Los Angeles, Ca., 90014, Office phone = 1-213-689-7763, Fax= 1-213-689-7765, e-mail = s1pts@aol.com, www.s1gard.com
10. Currently the S-1 GARD is the only device being sold to eliminate the accidents that occur at large vehicle wheel wells.
11. I have personally reconstructed accidents on garbage trucks operating in high-density population neighborhoods. The Forensic Engineering evaluation of these accidents did show that the victims where impacted at thel well of the rotating rear wheel.

Chapter 32

Bike Fit

Introduction

In determining the causal factors of bicycle accidents, the issue of bike fit is often overlooked. Often, an improper bike fit can cause an incredibly unstable relationship between the cycle and the cyclist. There are many studies available delineating the optimal weight distribution of a cyclist on a road or mountain bike. Weight distribution is defined as the amount of weight supported by each axle. One of the methods for measuring weight distribution is to place simple bathroom scales under the front and back wheels. The rider can then sit on the seat of the cycle with his or her hands on the handlebar hoods. The total weight of the cycle and rider can then be measured by reading the weight dial on the bathroom scales.

Typically, for a general use cycle, the total weight distribution should be 60 percent on the rear wheel and 40 percent on the front wheel.

Analysis of General Methods of Bicycle Fit

In order to fit a bicycle safely to the user, the following definitions are typically used by most retailers.

Frame Measurement

The most common method of measuring a bike frame is from the center of the bottom bracket to the top of the seat tube lug. This method is usually determined in centimeters. For example, a bike that is referred to as a 54 cm, usually means the distance from the center of the bottom bracket to the top of the seat tube lug is 54 cm. Occasionally, one may use a center to center measurement. This refers to the distance from the center of the bottom bracket to the center of the seat tube lug. Center to the top of the seat tube is typically used in the retail industry.

Center to center measurements are typically 1.0-2.0 cm smaller than center to top measurements.

Frame sizes for typical off the shelf manufactured bike frames are usually in the 49-60 cm center to top range.

Cyclist Critical Measurements

There are many methods for optimizing power output on a bicycle. The transmittal of power to pedals directly equates to how fast the cycle will travel. Different frames are available with seat tube angles varying from 73-75 degrees. The seat tube angle is defined as the angle from a line passing horizontally through the hubs and the seat tube. The seat tube angle allows the general user to position their knees over the pedal axle. This allows the retailer to position the saddle fore or aft so that the knee can be placed over the pedal axle.

Various positions are utilized by the racing community to vary power to the bike frame. For example, someone who is a time trialer may choose a seat angle greater than 75 degrees. Smaller wheels, such as 650 cc instead of 700 cc wheels, can be used to aid in acceleration. These various ramifications should only be accomplished by very experienced cyclists and are not considered pertinent to this chapter on general bike fit.

The critical measurements of an individual that will be using a standard road or mountain bike, are installed to distribute the weight in the 60/40 ratio described previously.

The basic method for determining seat height is to first measure inseam length. This should be done by standing in stocking feet on a flat hard surface and snugging a carpenters level into your crotch. With the carpenter's level parallel to the floor, the distance from the top of the level to the floor should be determined. This is the inseam length that will be used for determining frame size. The inseam measurement should be multiplied by 0.883. The product of the two numbers is the distance from the center of the bottom bracket to the top of the saddle. To determine bike frame size, the formula is:

inseam * 0.67 = center of bottom bracket to the top of the seat lug.

Either centimeters or inches can be used in this analysis.

Once the seat height is known, the top tube length can be determined. That is accomplished by dividing the cyclist's height by the inseam length. This ratio will yield the length of the top tube by using the following table.

Table 1: Size Ranges as a Function of the Seat Tube and Top Tube Length

Ratio	Torso	Size
2.0-2.2	Average	Top Tube = Seat Tube
> 2.2	Long Torso	Longer Top Tube
< 2.0	Short Torso	Shorter Top Tube

The ratio of the top tube to the seat tube varies from manufacturer to manufacturer. Top tubes tend to be within two centimeters of the seat tube length.

<u>Example Calculation</u> - Assume a person's inseam length is 31.0 inches and their height is 5 feet 11 inches.

$$\text{Ratio} \quad \frac{71.0}{31.0} = 2.29$$

$$\text{Seat Height} = 31.0 * 0.883 = 27.37 \text{ inches} = 70 \text{ cm}$$

$$\text{Frame Size} = 31.0 * 0.67 = 20.8 \text{ inches} = 52.8 \text{ cm}$$

A 53 cm frame would yield 17 cm to the top of the seat.

The top tube would be 54-56 cm due to the ratio of 2.29 revealing a long torso.

To be safe, the entity sizing the bike to this individual should insure that the top tube is long enough to obtain proper weight distribution.

Seat Adjustment

Once the seat tube and top tube measurements are defined, and a frame chosen, the rider needs to be positioned on the cycle. With the cyclist on the bike, the rider's feet should be placed on the pedals at the 9 o'clock and 3 o'clock position. A plumb line should be attached to the front of the knee and should just touch the knee and align with the end of the crank arm. The seat can be adjusted fore, aft, up or down to insure this relationship exists.

Reach Adjustment

Reach is defined as the distance from the nose of the saddle to the transverse centerline of the handlebar. After step 3 is completed, and the seat is level, the rider should be placed on the cycle with the hands on the drops on a road bike or the grips on a mountain bike. A plumb line from the riders nose with the head bent 45 degrees should bisect the handlebar stem. Another rule of thumb would be for the rider to look at the front axle. The handlebar should block the line of sight to the axle if the handlebar reach is correct.

Conclusion

The four steps defined herein are the minimum procedures that should be taken at the point of sale. If the Forensic Engineer is evaluating bike fit as a potential causal factor of an accident, then checking the aforementioned four steps is a good method for determining the handling characteristics of the machine. If the weight of the rider is distributed too far over the front axle, then it is difficult to steer the bicycle. It is also extremely easy for rotation over the front to occur upon application of the front brake. (See the Chapter "The Engineering Dynamics of a Cyclist Being Thrown over the Front of a Cycle During a Sudden Stop.")

There are other items for a final bike fit, such as proper pedal adjustment and seat adjustment fore and aft. These issues deal with power transmission and are not the driving force in weight distribution.

Bibliography

1. Bernhardt, Gale. "The Female Cyclist," Velo Press, 1830 n 55th Street, Boulder, Colorado, 80301, 1999.
2. Fit Stick, Cyclemetrics, www.cyclemetrics.com.
3. Talbot, Richard, P.E. "Designing and Building Your Own Frameset," Second Edition, The Manet Guild, 1984.
4. Van der Plas, Rob. "Bicycle Technology," Bicycle Books, 1282 - 7th Avenue, San Francisco, CA 94941, 1995.

Figure 1: Component Identification for Bike Fit

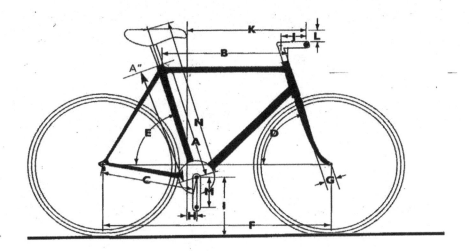

A'—Seat tube length, measured center-to-center
A"—Seat tube length, measured center-to-top
B—Top tube length
C—Chainstay length
D—Head tube angle
E—Seat tube angle
F—Wheel base
G—Fork offset
H—Seat setback
I—Bottom bracket height
J—Stem length
K—Saddle to transverse centerline of handlebar
L—Seat to handlebar drop
M—Crank

Chapter 33

Bicycle Wheel Performance[1]

By Kraig Willett

Sir Isaac Newton proposed his second law of motion nearly 350 years ago. This law elegantly describes the behavior of many systems, bicycles included. It is difficult to argue with the analytical results of this equation. It is my goal to use Newton's law to demonstrate what variables determine wheel performance and their order of significance. Furthermore, I will show how wheels rank in the big picture of overall cycling performance.

Newton declared that the sum of all the forces on a system (F) are equal to its mass (m) times its acceleration (a), or:

$$F = ma$$

Several companies offer power-measuring devices to help you quantify cycling performance. Power is defined as how much work is done during a given period of time, and it is a more convenient variable for illustrating wheel performance effects quantitatively. We can easily manipulate Newton's law to incorporate this rider power variable. Our once simple equation now becomes:

$$\text{Power} = \text{Velocity}^* (\text{Drag} + \text{Inertial} + \text{Rolling})_{\text{Force}} + \frac{(\text{Gravity})_{\text{work}} + (\text{Misc})_{\text{work}}}{\Delta \text{time}}$$

- Aerodynamic drag resulting from
  - Rider, frame, front wheel, rear wheel
- Inertial forces
  - Bike and rider mass
  - Front and rear wheel mass and its distribution (wheel inertia)
- Rolling resistance
  - Total system mass
  - Tire pressure and width
- Gravitational forces

- Bike and rider mass, front and rear wheel mass
- Miscellaneous
  - Drive train losses
  - Component flex

The details of this equation are beyond the scope here, so I will spare you the agonizing math (see Appendix A if that kind of stuff cranks your chain). The bulleted variables above are the only ones that matter when determining wheel performance. The question then becomes, which is the most important?

The obvious answer is that it's the rider pushing the pedals that matters most. This may not be the answer you were looking for, but swapping out your wheels will not make you the next Eddy Merckx. They may, however, help out a bit in your next race.

Based on the model developed in Appendix A, and data I have collected while riding (velocity and elevation profiles), it is possible to quantify the effect your wheels have on the average power required to complete a given ride/race. I will look at the following three course examples:

1. 6.5 hour solo training ride with 1200m/3940ft of climbing (31 kph/19 mph average speed from El Cajon, CA to Mexicali, B.C. Mexico)

See Appendix B for more details.

2. Uphill portion only of training ride above (similar to an uphill TT – 27 kph/16.8 mph average speed)

See Appendix C.

3. 1999 Barrio Logan Grand Prix – 50 minute Pro 1,2 criterium with 10 m elevation change per lap (45 kph/28mph average speed – sitting in the pack)

See Appendix D.

With my model, I varied the wheel mass, wheel inertia, and wheel aerodynamic variables independently to come up with the data in the following table:

Table 1:

	Training ride Avg power (watts)	Training- Uphill portion only Avg Power (watts)	Pack-Filler Criterium Avg Power (watts)
Reference Case (32 hole wire spoked wheel with 2.0/1.5mm spokes)	193.66	237.2	323.33
50% Lower Front Wheel Mass (all carbon rim wheels	193.38 (.14% decrease)	236.3 (.38% decrease)	322.70 (0.2% decrease)
50% Lower Rear Wheel Mass	193.30 (19% decrease)	236.1 (.46% decrease)	322.52 (.25% decrease)
50% Lower Front Wheel Inertia	193.63 (.015% decrease)	237.19 (.004% decrease)	323.27 (.019% decrease)
50 % Lower Rear Wheel Inertia	193.62 (.020% decrease)	237.18 (.008% decrease)	323.26 (.022% decrease)
50% Lower Front Wheel Drag Coefficient (deep section rim >40mm/1.57 in - Cosmic Carbone, Corima)	190.09 (1.8% decrease)	234.17 (1.1% decrease)	319.22 (1.3% decrease)
50% Lower Rear Wheel Drag Coefficient	192.49 (.6% decrease)	235.3 (.8% decrease)	320.25 (1% decrease)

So, what do all these numbers mean? It means that when evaluating wheel performance, **wheel aerodynamics are the most important, distantly followed by wheel mass**. Wheel inertia effects in all cases are so small that they are arguably insignificant.

How can it be that wheel inertial forces are nearly insignificant, when the advertisements say that inertia is so important? Quite simply, inertial forces are a function of acceleration. In bike racing this peak acceleration is about .1 to .2 g's and is generally only seen when beginning from an initial velocity of 0 (see criterium race data in Appendix D). Furthermore, the 0.3kg/0.66lb difference in wheels, even if this mass is out at the rim, is so small compared to your body mass that the differences in wheel inertia will be unperceivable. Any difference in acceleration due to bicycle wheels that is claimed by your riding buddies is primarily due to cognitive dissonance, or the placebo effect (they paid a lot of money for the wheels so there must be some perceivable gain).

The following table illustrates how other variables in the power equation affect overall performance.

Table 2:

	Training Ride Avg Power (watts)	Training - Uphill Portion Only Avg Power (watts)	Pack-Filler Criterium Avg Power (watts)
Reference Case (32 hole wire spoked wheel with 2.0/1.5 mm spokes)	193.66	237.2	323.33
4.5 kg/9.9 lb Lower Body Mass	190.2 (1.8% decrease)	229.5 (3.2% decrease)	315.1 (2.5% decrease)
50% Lower Rider Drag Coefficient	143.3 (26% decrease)	204.3 (13.9% decrease)	267.4 (17.3% decrease)
50% Lower Frame Drag Coeffiecient	186.4 (3.7% decrease)	232.1 (2.2% decrease)	315.0 (2.6% decrease)
50% Lower Coefficient of rolling Resistance	174.8 (9.7% decrease)	225.5 (4.9% decrease)	308.1 (4.7% decrease)

It can be seen that rider aerodynamics dominates the power requirements of racing bikes. Frame and combined wheel effects are roughly equivalent, and it is interesting to note how power requirements are affected by rolling resistance changes in the examples.

Roughly, the average rider power requirements on a course with a zero net elevation gain is broken down into 60% rider drag, 8% wheel drag, 8% frame drag, 12% rolling resistance .5% wheel inertia forces and 8% bike/rider inertia. The uphill TT example given is a special case where the rider aerodynamics and the bike/rider weight have nearly equal contributions to power – somewhere around 35% each with wheel mass contributing around 1%. The steeper the hill, the more important mass becomes and the less important aerodynamics becomes. In all cases, however, there is approximately 3% of the average power unaccounted for.

Drive train losses and flexing of bicycle components can be placed into the miscellaneous term of the power equation. Even though these flexural losses are miniscule when compared to wheel inertial power requirements, lateral stiffness/deflection of wheels has its place in a performance analysis. My requirements are rather simple: road wheels should not rub brake pads during sprints and out of the saddle climbing, provided there is 2mm/0.079in of pad/rim clearance on either side. For reference, 2mm is the clearance when your dual pivot brakes are opened up, yet they still have sufficient braking power available.

In summary, wheels account for almost 10% of the total power required to race your bike and the dominant factor in wheel performance is aerodynamics. Wheel mass is a second order effect (nearly 10 times less significant) and wheel inertia is a third order effect (nearly 100 times less significant). The best wheels in terms of performance are the ones that are lightweight, aerodynamic, don't rub brake pads and are strong enough to get you to the finish line. The problem with these high performance wheels, though, is that they sacrifice on the other two key variables important in wheel selection: durability and price. High performance wheels are neither durable nor cheap. Nothing is ever easy, is it?

Endnotes

1. This chapter was reprinted with the permission of Bike.com. the article was published on their web page on August 2, 2001. Bike.com is the pre-eminent web site for the transfer of solid technical information on cycling.

Appendix A:
Equation of Motion

$$Power = Velocity * [Drag + Inertial + Rolling]_{Force} + \frac{(Gravity)_{work} + (Misc)_{work}}{\Delta time}$$

Aerodynamic term:

$$V\left\{\left[\frac{1}{2}\rho V^2 C_d A_w\right]_{FrontWheel} + (1-protection)\left[\frac{1}{2}\rho V^2 C_d A_w\right]_{RearWheel} + \left[\frac{1}{2}\rho V^2 C_d A_w\right]_{Frame} + \left[\frac{1}{2}\rho V^2 C_d A_w\right]_{rider}\right\}$$

Where:

ρ = air density
V = velocity
C_d = drag coefficient
A = reference area (side area for wheels, frontal area for rider and frame)
Protection = reduction in C_d of rear wheel due to its proximity to frame = .25

Inertial term:

$$\omega[I\alpha]_{Frontwheel} + \omega[I\alpha]_{Rearwheel} + V \cdot M_{Tot} a$$

Converting to linear acceleration terms and discretizing we have:

$$\omega = \frac{V}{R}$$

$$\alpha = \frac{d\omega}{dt} = \frac{d}{dt}\left(\frac{V}{R}\right) = \frac{1}{R}\frac{\Delta V}{\Delta t}$$

$$a = \frac{\Delta V}{\Delta t}$$

$$\therefore$$

$$V\left\{\left[I\frac{1}{R^2}\frac{\Delta V}{\Delta t}\right]_{Frontwheel} + \left[I\frac{1}{R^2}\frac{\Delta V}{\Delta t}\right]_{Rearwheel} + M_{Tot}\frac{\Delta V}{\Delta t}\right\}$$

Appendix B:

6.5 hour solo training ride from El Cajon, CA to Mexicali B.C., Mexico

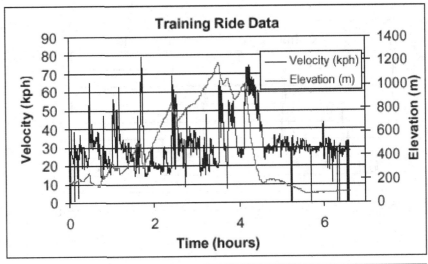

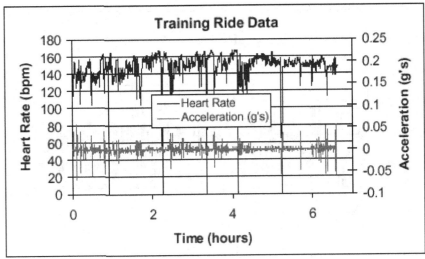

Appendix C:
Uphill portion of training ride

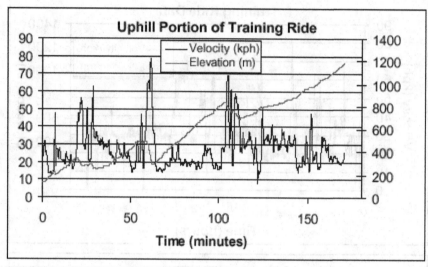

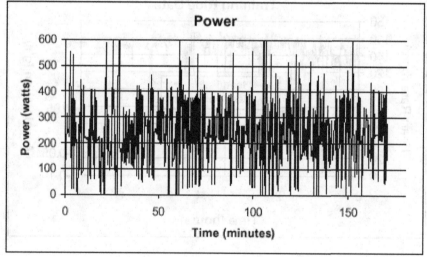

Appendix D:

1999 Pro 1,2 Barrio Logan Grand Prix Criterium (sat in the whole time)

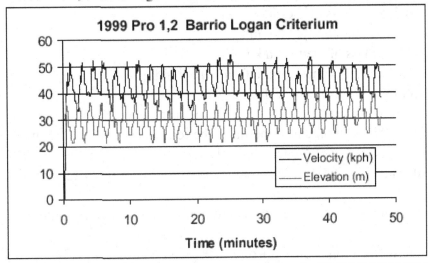

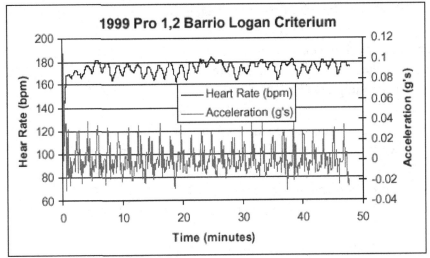

Reference case (annotated input parameters for a Matlab file):

mrider = 75; <= mass of the rider in kg
mbike = 7; <= mass of the bike in kg
mtire=.4; <= mass of the tire in kg
mfw = mtire + .7; <= mass of the front wheel in kg
mrw = mtire + .9; <= mass of the rear wheel in kg
Ifrontrim = .04; <= moment of inertia of the front wheel without tire in kg-m^2
Irearrim = .05; <= moment of inertia of the rear wheel without tire in kg-m^2
R = .3429; <= radius of the front wheel
Ifw = Ifrontrim + mtire * R^2; <= moment of inertia of f. wheel with tire in kg-m^2
Irw = Irearrim + mtire * R^2; <= moment of inertia of r. wheel with tire in kg-m^2
Mtot = mrider + mbike + mfw + mrw; <= total mass of system in kg
Crr = .004; <= coefficient of rolling resistance
g = 9.81; <= gravitational acceleration m/s^2
Cdrider = .5; <= drag coefficient of rider
Cdbike = .5; <= drag coefficient of bike
Cdrw = .04; <=drag coefficient of rear wheel
Cdfw = .04; <= drag coefficient of front wheel
Afw = pi * R^2; <= projected side area of front wheel
Arw = Afw; <= rear wheel area is equivalent to front wheel area
Arider = .4; <= projected frontal area of rider
Abike = .06; <= projected frontal area of bike
protection = .25; <= factor of reduction of rear wheel drag coefficient
temperature = 25; <= ambient temperature in Celsius
pressure = 101325; <= ambient pressure in Pa
temperature = temperature + 273.15; <= conversion to Kelvin
rho = (3.484e-3 * pressure)/(temperature); <= air density calculation kg/m^3

Chapter 34

Cyclist-Motor Vehicle Accidents

By Bob Mionske, Esquire

Introduction

This chapter is composed of annotations of legal cases that involve a cyclist, usually the plaintiff, and a motorist, usually the defendant, driving some kind of motor vehicle (e.g., car, truck, bus, tractor, motorcycle, etc.), that are involved in an accident. The basic facts are provided, as well as the trial court judgment, and the reviewing court (court of appeals or supreme court) decision. Typically, the annotation will include only that part of the decision dealing with a legal issue that is particular to cyclists. Some cases have limited precedential value as the statutes in these jurisdictions have changed. They are included for whatever precedence remains in the controlling jurisdiction and for their persuasive value in other jurisdictions.

Scope

Cyclists Overtaken by Motorists

The most frequent fact pattern in these compiled case annotations is where a motorist collides with a cyclist while attempting to pass the cyclist from behind. This type of accident occurs due to the motorist's negligence in failing to keep a proper lookout; failing to control his/her vehicle; failing to maintain a reasonable speed; failing to yield the right-of-way; following too closely; passing too closely (crowding); and/or passing on the right. These accidents happen because of the cyclist's negligence, as well. Examples of cyclists' negligence, in overtaking accidents, include: failure to ride in a straight line; turning without signaling; changing lanes without signaling; failure to equip his/her bicycle with a light (front and rear) while riding at night; failure to ride on the right; and riding two (or more) abreast (legal in some states and prohibited in others).

Intersection Accidents

The next most common fact pattern involves a cyclist and a motorist colliding at an intersection. These accidents occur in both controlled and uncontrolled intersections. Failure to yield, stop or signal a turn are common examples of motorist negligence. The cyclist is often guilty of the same offenses, and this fault lowers or forecloses his/her ability to recover damages.

Wrong-Way Cyclists

Motorist/cyclist accidents are often unavoidable when cyclists ride in the wrong direction, against traffic. Legal liability in these cases depends on, among other things, jurisdictional law, where the cyclist is riding (berm, shoulder lane, etc.), and whether the cyclist was a legal minor.

Sidewalk, Crosswalk Riding

Many cases describe situations where the cyclist is riding on sidewalks and crosswalks and is hit by a motorist. The motorist may be liable for failure to yield under (sidewalk use statutes), and the cyclist may be liable for the accident by riding on a sidewalk or crosswalk in prohibition of state or local law.

Entering or Crossing Roadways

Other annotations involve circumstances where a cyclist or motorist crosses or enters a roadway from a shoulder or private drive, failing to yield to favored traffic. Typical examples include a cyclist attempting to ride across a highway and a motorist backing out of a driveway.

Vehicle Door and Unsafe Loads

Lastly, the reader will find descriptions of legal cases brought in negligence where a parked motorist opens his/her door, and a cyclist rides into it. The vehicle is usually parked, and the accident occurs when the cyclist is riding past and is injured when the motorist or passenger opens a door. Unsafe load cases occur most commonly where an overtaking motorist has objects loaded on or about his/her vehicle, extending beyond the borders of his/her vehicle and strikes a cyclist when passing.

In the research of this chapter, an attempt was made to include all cycling/motorist cases; however, there may be some cases not listed that are of value to a researcher on bicycle/car cases. Therefore, it is advised that this resource be used merely as a starting point and as a quick guide in obtaining some leading cases in a particular jurisdiction. Further, a great effort was made to be accurate as to the facts and procedure of the court proceedings. Again, these annotations are a starting point for any legal research, and the complete case should be obtained and reviewed in order to determine its precise facts, law, holding, and precedential value.

Motorists Cases

Alabama

A cyclist was riding at night without lights. An overtaking motorist did not see the cyclist until just before he came upon him. The motorist swerved into the opposing traffic lane and collided with another motorist. After the jury returned a verdict in favor of the cyclist, the trial court judge granted the motorist's motion for a new trial. The Supreme Court affirmed. *Clark v. Chitwood*, 339 So. 2d 1017 (Ala. 1976).

An eight-year-old cyclist, riding at night, was hit by an overtaking motorist. The Supreme Court held that the evidence of whether the cyclist's bicycle was equipped with a headlight was properly allowed where a headlight may have alerted the motorist of the cyclist's presence on the road, irrespective of the fact that the cyclist was hit from the rear. The jury should not have been instructed on the cyclist's contributory negligence. *McWhorter v. Clark*, 342 So. 2d 903 (Ala. 1977).

Arizona

A minor cyclist, aged ten, riding his bicycle in the crosswalk, was struck by a motorist who was making a left-hand turn on a green light. The Arizona Supreme Court held that the trial court had properly refused the defense requested instruction on the statute requiring that bicycles be ridden on the right hand side of the street, since it was immaterial to whether he was negligent while riding his bicycle in the crosswalk. *Maxwell v. Gossett*, 612 P.2d 1061 (Ariz. 1980).

A sixteen-year-old cyclist collided with the stationary police car and after being unable to remove his foot from the toe clip was injured when he fell to the roadway. The cyclist had been riding the wrong way on a one-way street when the accident occurred. The trial court, citing *Maxwell v. Gossett*, 612 P.2d 1061 (Ariz. 1980) [see above annotation], instructed the jury that the cyclist had not violated traffic law, as a matter of law, by riding his bicycle the wrong way on a one-way street. The appellate court reversed and held that cyclist was precluded from claiming compensation from county due to his violation of statue prohibiting wrong way travel, despite cyclist's argument that statute only applied to motorists. *Rosenthal v. County of Pima*, 791 P.2d 365 (Ariz. 1990).

Arkansas

Two twelve-year-old cyclists were riding on a road at night. An overtaking motorist hit and killed them. The jury returned a verdict in favor of the cyclist, and the trial court judge entered the judgment accordingly. The Supreme Court affirmed, holding that the trial court did not error in denying the requested instruction on unavoidable accident. Proper lookout was not established, as a matter of law, on the part of the motorist by evidence that the cyclist's had violated the bicycle light statute. *Norman v. Gray*, 383 S.W.2d 489 (Ark. 1964).

A cyclist was riding early on a January morning. An overtaking taxicab hit him. The jury returned a verdict for the taxicab driver. The trial court judge granted the motion for a new trial. The Supreme Court reversed, holding that it was error to grant a new trial where the jury could reasonably find the cyclist negligent in spite of the cyclist's wearing of reflective clothing and bicycle reflectors. The jury's verdict was not clearly against the preponderance of the evidence. *Razorback Cab v. Martin*, 856 S.W.2d 2 (Ark. 1993).

California

A fifteen-year-old was riding as a passenger on a bicycle operated by another boy. They were hit by an intoxicated motorist. There were several motorists involved in the accident. The jury returned a verdict in favor of the defendants. The Court of Appeals affirmed, holding, inter alia, that any contributory negligence on the part of the operator of the bicycle could not be imputed to the passenger. *Linzsey v. Delgado*, 54 Cal. Rptr. 762 (Cal. Ct. App. 1966).

A city bus struck a mentally impaired thirty-three-year-old bike messenger. The lower court entered a judgment in favor of the cyclist, and the municipality appealed. The Court of Appeals held that individuals using the highways have the right to expect that other members of the traveling public will obey traffic laws and exercise the adult standard of ordinary care. The court stated that it had been erroneous for the lower court to give an instruction on the impaired facilities of the bicyclist in the absence of any evidence that the impaired facilities of the bicyclist played a role in causing the collision. *Fox v. City and County of San Francisco*, 120 Cal. Rptr. 779 (Cal. Ct. App. 1975).

Colorado

A motorist struck and killed a cyclist at an intersection. The motorist testified that he never saw the cyclist prior to hitting him. An eyewitness testified that the cyclist appeared to have been attempting to, or did, stop his bicycle and was standing with both feet on the ground. The court affirmed the trial court's award to plaintiff, stating that the evidence warranted the jury's inference that had the motorist exercised ordinary care in looking to the left and right prior to entering the intersection, he would have seen the cyclist. Although the motorist was within the law with respect to the right of way, there was sufficient evidence in the record to warrant the jury's inference that the motorist breached his duty to use the required ordinary care and skill in the control of his vehicle in order to avoid an accident. Additionally, the court held that the cyclist was not contributorily negligent, as a matter of law, for failing to equip his bicycle with a horn, nor was he contributorily negligent, as a matter of law, for failing to yield the right of way to the motorist's automobile, which entered the intersection from his right, since the motorist himself had been negligent in the operation of his own vehicle. *Green v. Pedigo*, 75 Cal. App. 2d 300, 170 P.2d 999 (1946).

A fourteen-year-old cyclist and a thirteen-year-old cyclist were on a ride together. They came to an controlled intersection. They entered the left-turn lane and waited for a green arrow. Because they were unable to trigger the detector, they never activated the green arrow. After several minutes, they decided to make a left turn without the arrow. While their direction had a green light (but not a green left-turn arrow), they attempted the turn. The fourteen-year-old proceeded into the intersection. The first lane of opposing traffic, which also had a green light, stopped upon seeing the boys attempting to cross in front of them. A motorist, also in the inner lane, seeing the traffic stopped ahead, switched to the outer lane and accelerated. The motorist hit and killed the fourteen-year-old cyclist. The trial court judge directed a verdict in favor of the motorist. The Court of Appeals, Tenth Circuit, reversed. The court stated that, in borderline cases, a trial court judge must recognize the possibility that, whatever might be his own view, other fair-minded men might reasonably arrive at a contrary conclusion. The motorist's view of the cyclist was obstructed by the cars in the left lane. These cars did not move forward when the light changed. Under the circumstances, it is not unreasonable to infer that the motorist's act of accelerating when she entered the intersection was negligent. *Christopherson v. Humphrey*, 366 F.2d 323 (10th Cir. 1966).

A minor cyclist collided with an automobile when he rode his bicycle through an intersection cutting the corner in front of defendant motorist. The Supreme Court affirmed the lower court's holding in the motorist's favor, stating that the motorist did all that a reasonable person could do under the circumstances. The motorist was traveling at the posted speed; he stopped within six feet short of the intersection; and, although his vehicle was one foot over the centerline, skid marks showed he had applied his brakes on his side of the road. The court stated that a driver who suddenly realizes there is an impending collision because of another's negligence cannot be charged with negligence for errors in judgment when instantaneous action is required. *Virgil v. Kinney*, 441 P.2d 7 (Colo. 1968)

A fourteen-year-old boy, who was riding his bicycle with a friend, was struck by a vehicle attempting to overtake him. The lower court directed a verdict on the question of liability in favor of the plaintiff cyclist. The Supreme Court affirmed, stating that the defendant motorist's failure to adhere to the statutes regarding overtaking a vehicle and maintaining a safe distance between vehicles constituted negligence per se. The motorist admitted that the plaintiff cyclist and his friend (riding behind plaintiff on another bike) were within the extreme right lane and that he (motorist) was partly in the same lane as the boys. Additionally, the motorist admitted that there was no traffic in either the center or left lane that would have prevented his use of these lanes while passing the two cyclists. *Thompson v. Tartler*, 443 P.2d 365 (Colo. 1968).

A cyclist collided with a motorist's vehicle at a four-way intersection marked by "stop ahead" warning signs and four stop signs. The motorist admittedly failed to stop at the intersection, which she entered at a speed of approximately 25-35 miles per hour. Just prior to entering the intersection, the cyclist slowed down, looked left and right, and attempted to stop when he observed motorist's vehicle in the intersection. However, because of sand and gravel in the intersection, his bicycle skidded, and he was unable to stop in time to avoid the collision. The Supreme Court upheld the lower court's verdict on liability in favor of the plaintiff, stating that issues of negligence, contributory negligence, and the last clear chance doctrine were proper issues for the jury and denying the motorist's motions for dismissal, directed verdict, and judgment notwithstanding the verdict. *Kistler v. Halsey*, 481 P.2d 722 (Colo. 1971).

A cyclist and his passenger, were traveling downhill on a mountain highway when they were struck from behind by a motorist's vehicle. The motorist appealed the trial court's ruling in favor of the plaintiffs, stating that he did not have a better opportunity to avoid the accident. The Court of Appeals disagreed with the defendant motorist and affirmed the lower court's ruling in favor of the cyclists. The court held that the issue of whether the motorist had the last clear chance to avoid the accident had been properly put before the jury where the record supported the jury's determination that the cyclists were unaware of the motorist; that unless the motorist took evasive actions, the cyclists would be injured; and, that the motorist was aware of this position of peril. *Gomez v. Black*, 511 P.2d 531 (Colo. Ct. App. 1973).

Connecticut

A ten-year-old girl was riding as a passenger on a bicycle being operated by a twelve-year-old boy. An overtaking motorist collided with them as the boy veered left. The judgment was entered in favor of the ten-year-old plaintiff. The Supreme Court affirmed, holding that the negligence of the operator could not be imputed to the passenger. The statute prohibiting riding double was not in existence at the time of the accident, 1937. *Johnson v. Shattuck*, 3 A.2d 229 (Conn. 1938).

A cyclist moved to set aside a jury verdict in favor of the defendant motorist. The cyclist was riding his bicycle on a private road when a motorist made a left-hand turn in front of him without using his signal and without coming to a stop. In order to avoid hitting the motorist, the cyclist turned his bicycle away from the motorist's vehicle and was struck by another vehicle. The cyclist argued that the jury's finding of his negligence at seventy percent should be set aside because there was no evidence that he was speeding or operating his bicycle out of control. The Superior Court affirmed the jury's verdict, stating that it was supported by the evidence and therefore, the court did not have a duty to set it aside. *Niel v. Griswold*, 1992 WL 172138 (Conn. Super. Ct.).

A cyclist was riding south against traffic. A motorist, heading north, swerved to avoid the cyclist, but hit the cyclist with the passenger side mirror of his vehicle. The trial court judge directed verdict for the motorist. The Appellate Court reversed, holding that a jury question was presented as to whether the motorist's operation of his vehicle was a substantial factor in the cyclist's injuries. Inference that the accident would not have occurred had the motorist maintained a proper lookout and kept his vehicle under proper control would have been both reasonable and logical in light of the evidence of clear visibility and fair weather. *Phinney v. Casale*, 671 A.2d 851 (Conn. App. Ct. 1996).

Delaware

A cyclist, on a group ride, was riding in a lane reserved for bicycles. When the cyclist entered an intersection, a motorist turned left into the cyclist. After the jury verdict, the trial court judge entered judgment for the motorist. The Supreme Court held that the jury's verdict was manifestly against the weight of the evidence. The cyclist was heading south in a lane reserved for cyclists, was struck broadside by a northbound motorist who made a left turn in front of the cyclist, who had the right-of-way. The motorist had a duty to remain stationary and yield the right-of-way before turning. The motorist's failure to do so constituted negligence per se. *Luskin v. Stampone*, 386 A.2d 1137 (Del. 1978).

District of Columbia

An eleven-year-old cyclist was riding eastbound and turned left at an intersection in front of a westbound motorist. The jury returned a verdict for the motorist. The cyclist moved for a new trial, and the motion was granted. The reviewing court held that by instructing the jury that the negligence per se doctrine applied to the minor cyclist, the trial court, in effect, charged the jury that the minor cyclist, as a matter of law, was capable of exercising the same degree of care as an adult. Such instruction was error. A violation of a traffic regulation by a minor cyclist is not negligence per se. It is merely one factor to consider in determining whether the minor cyclist was guilty of contributory negligence. The jury should consider whether the minor cyclist exercised reasonable care in light of his age, education, training, and experience. *Herrell v. Pimsler*, 307 F. Supp. 1166 (D.C. 1969).

A cyclist was riding in the right lane of three southbound lanes. A taxicab motorist turned right from the middle lane and collided with the cyclist. The jury returned a verdict in favor of the cyclist. The trial court judge set it aside based on the cyclist's contributory negligence. The Court of Appeals reversed, holding that the evidence at trial did not establish contributory negligence as a matter of law. *Williams v. Anderson*, 485 A.2d 198 (D.C. 1984).

Florida

A minor cyclist was riding with her two brothers, who were also on bikes, in single file. As an overtaking motorist passed, the cyclist fell to the roadway and was injured. The trial court entered judgment for the motorist after the jury verdict in the motorist's favor. The Court of Appeals reversed, holding that the instruction given to the jury on contributory negligence was error. Even though the motorist contended that evidence that the cyclist was holding a paper bag that was wrapped around his handlebars contributed to the accident, it was for the jury to decide if carrying such a bag was negligent and caused diminished control. *Lynn v. Pulford*, 200 So. 2d 201 (Fla. Dist. Ct. App. 1967).

A twelve-year-old rode his bicycle through a stop sign and collided with the rear portion of a tractor-trailer. The trial court granted summary judgment in favor of the motorist. The Court of Appeals reversed, holding that a question of fact existed as to whether the motorist was guilty of gross negligence in driving a fifty-five foot long tractor-trailer past the intersection marked as a school crossing at an excessive speed at a time when children were visible in the vicinity. *Youngblood v. Bowman Transportation, Inc.*, 261 So. 2d 206 (Fla. Dist. Ct. App. 1972).

A nine-year-old cyclist entered the roadway from the sidewalk on a rainy day and was hit by a motorist. The trial court judge entered a verdict for the motorist. The Court of Appeals reversed, holding that questions of comparative negligence were for the jury. There were conflicts and inferences in the evidence, such as whether the motorist applied his brakes soon enough, whether he was sufficiently observant, and whether he acted as a reasonably prudent person would have under the circumstances. *Goodstein v. Gary Fronrath Chevrolet, Inc.*, 458 So. 2d 869 (Fla. Dist. Ct. App. 1984).

A twelve-year-old cyclist was hit and killed by an overtaking multi-axle tractor-trailer. The jury returned a finding of no liability on the part of the motorist, and the trail court judge entered the judgment accordingly. The Court of Appeals reversed. The trial court's failure to give the requested instruction concerning violations of uniform traffic control law sections relating to unlawful speed and special hazards were reversible error. The jury was entitled to guidance on the statutes and the effect of a violation would have on its deliberations. The jury could reasonably infer that the failure to decrease speed was a violation of the statute. The motorist could have moved further over in his lane to avoid special hazard. *Sotuyo v. Williams*, 587 So. 2d 612 (Fla. Dist. Ct. App. 1991).

A cyclist and her friend were riding their bicycles on a highway in Jacksonville. An overtaking motorist whose vehicle was loaded with eight-foot timbers protruding, by two to three feet, out of the passenger-side windows attempted to pass, and the cyclist was struck by the timbers. The motorist and an employee of the Home Depot Store, where they had been purchased, had loaded the timbers into the vehicle. The trial court directed a verdict in favor of the store. The Court of Appeals reversed, holding that there was a jury question as to whether the negligence of the store in loading the timbers into the motorist's car was the proximate cause of the accident in which the cyclist was struck by the protruding timber. The jury could have reasonably found that the motorist's poor judgment in driving his automobile was not so unusual or bizarre that it could not have reasonably been foreseen. Therefore, the store from which the landscape timbers had been purchased, owed a duty to the cyclist struck by the timber protruding from the motorist's vehicle to safely load the timbers. It was not necessary that the store have foreseen the cyclist's presence on the road or manner in which her injuries occurred, since she was within the "zone of risk." *Kowkabany v. Home Depot, Inc.*, 606 So. 2d 716 (Fla. Dist. Ct. App. 1992).

A cyclist attempted to cross a road and was hit by a motorcyclist. The jury returned a verdict in favor of the motorcyclist, and a judgment was entered accordingly. The Court of Appeals affirmed, holding that the bicyclist's blood alcohol level was properly admitted under a hearsay exception for reports in the ordinary course of business despite cautionary language on the report that it was for "medical purposes only." *Andres v. Gilberti*, 592 So. 2d 1250 (Fla. Dist. Ct. App. 1992).

A cyclist was riding on a sidewalk when a motorist turned into a driveway and struck him. The trial court entered judgment in favor of the motorist. The Court of Appeals reversed, holding that it was an error not to give the requested instructions encompassing the statute requiring drivers of vehicles to avoid colliding with pedestrians or human-powered vehicles and giving them the same rights and duties as pedestrians. *Popejoy v. Harrison*, 615 So. 2d 823 (Fla. Dist. Ct. App. 1993).

Georgia

A twelve-year-old boy was riding as a passenger on the handlebars of a fourteen-year-old cyclist's bicycle. An automobile collided with them. The jury returned a verdict in favor of the motorist. The trial court judge granted a new trial. The Court of Appeals affirmed, holding that the instruction stating that the passenger must exercise due care for his/her own safety and cannot close his/her eyes to known or obvious dangers and treat himself/herself as dead freight was reversible error. The twelve-year-old had told the bicycle operator that a car was approaching and that the bicycle should be ridden out of the way. *Bennett v. George*, 125 S.E.2d 122 (Ga. Ct. App. 1962).

A cyclist was riding on the shoulder of a road and suddenly turned to his left in a diagonal path. An overtaking taxicab collided with him. The trial court entered judgment for the cyclist after the jury returned a verdict in his favor. The Court of Appeals affirmed, holding that a jury question was presented by evidence, and although a close question, the verdict was supported by some competent evidence. The instruction given concerning the requirement of a motorist to keep a lookout and failure to blow his horn were not in error. *Hughes v. Brown*, 143 S.E.2d 30 (Ga. Ct. App. 1965).

A cyclist was riding downhill on an unlit street. A motorist, with headlights turned on, was heading up the hill. The motorist turned left in front of the cyclist, and the two collided. The jury returned a verdict in favor of the cyclist, and the judgment was entered accordingly. The Court of Appeals reversed, holding that the jury instruction concerning violation of the statute requiring all motor vehicles to have adequate lights was prejudicially erroneous where there was no evidence presented that the motorist's vehicle lights were in violation of the statute. Evidence at trial showed that the lights were in working order and on low beam when the accident occurred. *Ingram v. Jackson*, 265 S.E.2d 29 (Ga. Ct. App. 1980).

A nine-year-old cyclist and a friend walked their bikes up a steep driveway. The friend rode down and into the road uneventfully. The nine-year-old then rode down the driveway and collided with a motorist. The trial court judge granted a directed verdict for the motorist. The Court of Appeals affirmed, holding that the motorist could not be held liable absence a showing of negligence. The motorist was traveling at fifteen miles per hour and came to a complete stop before the cyclist ran into her car. *Neal v. Miller*, 390 S.E.2d 125 (Ga. Ct. App. 1990).

A seven-year-old cyclist rode a ten-speed bicycle equipped with hand brakes down a driveway and out into the street where defendant motorist was passing. The child ran his bike into the passenger side door of the motorist's vehicle and was seriously injured. At trial the motorist moved for summary judgment, which was denied. The motorist appealed. The Court of Appeals held that the motorist was not negligent. He was traveling within the speed limit, moved to the center of the road when he noticed the children playing and had no way of knowing that the child would ride into the street. *Bell v. Leatherwood*, 425 S.E.2d 679 (Ga. Ct. App. 1992).

A ten-year-old cyclist attempted to cross a highway and collided with a motorist. Based on a jury verdict, the trial court entered a judgment in favor of the motorist. The Court of Appeals reversed, holding that it was an error for the trial court to instruct on negligence per se. In Georgia, children under the age of thirteen are not capable of negligence per se. The court also stated that the cyclist's may be found contributorily negligent depending on his mental and physical capabilities. *Sorrells v. Miller*, 462 S.E.2d 793 (Ga. Ct. App. 1995).

Idaho

A fourteen-year-old cyclist was riding on a narrow shoulder. A motorist approached the cyclist from the rear, and when he was approximately fifty feet away, the cyclist veered left into the path of the motorist and was hit and killed. The trial court entered the judgment on the jury verdict in favor of the cyclist. The Supreme Court affirmed, holding that there was a reasonable inference that the motorist, without appreciably slowing or giving an audible warning, approached the cyclist, crowded the right edge of the pavement, and the cyclist lost control of his bike. There was no evidence that the cyclist was contributorily negligent as a matter of law. A disinterested witness who was following the motorist at a distance of about three hundred feet testified that the motorist was driving close to the shoulder and never saw the motorist turn away from the cyclist or slacken his speed before hitting the cyclist. *Kelley v. Bruch*, 415 P.2d 693 (Idaho 1966).

An eight-year-old cyclist rode into the street and was hit by a motorist. The trial court judge entered a judgment for the cyclist after the jury returned a verdict in his favor. The Supreme Court affirmed, holding that the cyclist's alleged violation of the statute in riding his bicycle into the street without yielding the right-of-way to the motorist did not constitute contributory negligence as a matter of law. The question of contributory negligence was for the jury. *Davis v. Bushnell*, 465 P.2d 652 (Idaho 1970).

An eight-year-old cyclist was riding down a country road. He was hit by a motorist attempting to overtake him. After a jury verdict in favor of the cyclist, the trial court entered judgment accordingly. The Supreme Court affirmed, holding that even if the westbound automobile, which struck the westbound cyclist, had not crossed grooves approximately thirty inches form the edge of the road, such would not have warranted an inference that the cyclist had not operated his bicycle as near to the right-hand side of the pavement as practicable, and thus had been negligent. The motorist had testified that she was in no position to state where the cyclist was in the road, as she never saw him before impact. The requested inference that the cyclist was in the roadway required speculation and conjecture and was properly rejected by the trial court judge. *Owen v. Burcham*, 599 P.2d 1012 (Idaho 1979).

Illinois

A fourteen-year-old cyclist was riding down a graveled country road. He testified that he heard the horn from a motorist's automobile approaching from behind and moved to the extreme right side of the road and was hit. The motorist testified that she had to sound her horn twice before the cyclist turned to look, that the cyclist was riding on the left side of the road and veered to the right just as she was passing and that she could not avoid hitting him. The trial court entered judgment on the jury's verdict for the cyclist. The Court of Appeals affirmed, holding that the trial court properly refused to give the instruction holding the child to the same standard as an adult and properly instructed on the failure to yield the right-of-way, operating an automobile at high rate of speed considering the circumstances, and failure to keep a proper lookout. *Conway v. Tamborini*, 215 N.E.2d 303 (Ill. App. Ct. 1966).

An eleven-year-old cyclist heading north as he made a left-hand turn to enter a parkway and was hit by a motorist heading south. The jury returned a verdict in favor of the motorist. The trial court judge granted a new trial based on an erroneous jury instruction on contributory willful or wanton conduct. The Court of Appeals reversed, holding that an eleven-year-old boy could be guilty of contributory willful and wanton conduct based on the standard of a reasonably careful person of the same age, intelligence, and experience. Crossing the southbound lane, before having made any effort to observe oncoming traffic, raised an issue for the jury. *Wegler v. Luebke*, 231 N.E.2d 109 (Ill. App. Ct. 1967).

A ten-year-old cyclist collided with a camper truck. The trial court entered a judgment in favor of the cyclist. The Court of Appeals affirmed that the evidence generated a jury question as to whether the motorist was negligent in failing to pass to the left of the minor at a safe distance and in failing to maintain a proper lookout while cresting the hill before the collision. *Smith v. Stopher*, 261 N.E.2d 16 (Ill. App. Ct. 1970).

A cyclist was hit by a motorist while the cyclist was in the crosswalk. The trial court entered a judgment in favor of the cyclist. The Court of Appeals affirmed, holding that the verdict was not against the great weight of evidence when testimony showed that the motorist waited for the green light, turned left, cut the corner very sharply, and collided with the cyclist. There was ample evidence to sustain the jury verdict. *Zadura v. Debish*, 284 N.E.2d 28 (Ill. App. Ct. 1972).

Two twelve-year-old girls, riding double, entered a roadway and were hit by a motorist. The trial court entered judgment on a jury verdict for the motorist. The Court of Appeals affirmed, holding that the question of the passenger's contributory negligence was properly submitted to the jury. The trial court properly refused the cyclist's requested instruction, which stated that the issue of contributory negligence does not apply to the passenger. *Bertagnolli v. Ambler*, 295 N.E.2d 279 (Ill. App. Ct. 1973).

A nine-year-old cyclist, riding on a sidewalk, was hit by a motorist who was turning into an alley. The jury answered the special interrogatory finding the cyclist contributorily negligent. The Court of Appeals reversed, holding that the cyclist was riding in a normal manner on the sidewalk, that he did not ride fast, and that he looked both ways before entering the alley, and therefore, he was not contributorily negligent. The cyclist rode with an exceedingly high degree of care. The motorist testified that he never looked right before turning into the alley and that after he made his turn he was looking away from the direction of the young cyclist. *Zygadlo v. McCarthy*, 308 N.E.2d 167 (Ill. App. Ct. 1974).

A five-year-old cyclist rode out of a private driveway into the roadway and was struck by a motorist. The trial court entered judgment in favor of the motorist. The Court of Appeals affirmed, holding that the reconstruction expert testimony was admissible, even though it served to bolster the motorist's approximation of his speed, where the evidence responded to one of the main thrusts of the cyclist's case, and the testimony was generally consistent with the testimony of the cyclist's own testimony. There were no available eyewitnesses other than the motorist; therefore, a strong case was presented for admission of the reconstruction expert's testimony. *Coffey v. Hancock*, 461 N.E.2d 64 (Ill. App. Ct. 1984).

A fifteen-year-old cyclist and a motorist collided at an intersection. They both testified that they had a green light. The judgment was entered, finding the cyclist eighty percent at fault and the motorist twenty percent at fault. The Court of Appeals reversed, holding that the failure to submit the minor standard of care instruction was reversible error. There was no basis for the exception to the minor standard of care. *King v. Casad*, 461 N.E.2d 685 (Ill. App. Ct. 1984).

A nine-year-old cyclist was riding down a sidewalk when he collided with a motorist exiting an alley. After a bench trial, the trial court judge entered a judgment in favor of the motorist. The Court of Appeals affirmed, holding that the cyclist was not a "pedestrian" for purpose of the provisions of the Vehicle Code and Chicago Municipal Code, stating that the driver of a vehicle emerging from an alley shall yield the right-of-way to pedestrians, and therefore, the motorist alleged failure to yield the right-of-way was not prima facie evidence of negligence. The court explained that its review was limited to the propriety of the lower court's decision, not its own reasoning. The judgment was not against the manifest weight of the evidence, and the record does not mandate a reversal. *Bekele v. Ngo*, 603 N.E.2d 735 (Ill. App. Ct. 1992).

A truck driver waited at an intersection as a minor cyclist rode through the crosswalk and entered the sidewalk in front of him. The truck driver then turned left, proceeded 250-300 feet, and turned right into a parking lot. He heard an observer yell for him to stop, but he had already run over and killed the girl before he could. The motorist moved for summary judgment, which the U.S. District Court denied. Genuine issues of material fact existed as to whether the driver should have expected that a minor cyclist might ride on a sidewalk toward the parking lot driveway where the driver intended to turn, as to whether the driver failed to keep a proper lookout for the minor cyclist as he made the turn into the driveway, and as to whether the failure to keep a proper lookout was a proximate cause of the minor cyclist's death. A special or higher degree of diligence is required by Illinois law when children are known or should be known to be present. *Choe v. Ashdown*, 808 F. Supp. 1342 (N.D. Ill. 1992).

Two fifteen-year-old cyclists were riding on the same bicycle. One was seated, and the other had a hold of the handlebars while she pedaled. They came to an intersection, saw headlights approaching but thought they could cross in time. As they approach the centerline, they were hit. The trial court judge entered a judgment on a jury verdict in favor of the motorist. The Court of Appeals affirmed, holding that the evidence supported a jury verdict that the motorist had not been negligent in hitting the bicycle entering the intersection at the right angle. The cyclist had been riding double on a single bicycle in violation of a safety statute, and the operator of the bicycle admitted a lack of knowledge in shifting the gears of the bicycle and had testified to having a problem in getting started from a stopped position. *Hays v. Fabian*, 595 N.E.2d 162 (Ill. App. Ct. 1992).

A six-year-old cyclist informed her mother that she was going to ride down the driveway to the mailbox, which was located approximately one hundred and fifty yards from the house. A neighbor, returning to his farm, driving a tractor with a water wagon attached, followed by an attached fourteen-foot-wide field cultivator, swung to the left so that he could make a right at a "T" intersection. The neighbor was effectively blind in his left eye. After he completed his turn, he looked back and saw the young girl being dragged in the tines of his cultivator. The jury returned a verdict in favor of the motorist, and the trial court judge entered a judgment accordingly. The Court of Appeals affirmed and held that the evidence at trial provided support for the unavoidable accident theory. The evidence suggested that the young cyclist may have rammed into the farmer's water tank while he was making a turn at the intersection. Numerous witnesses testified that the view of the cyclist's approach was obstructed. *Syrcle v. Springer*, 606 N.E.2d 742 (Ill. App. Ct. 1992).

A ten-year-old cyclist and her friend were at a controlled intersection waiting to cross an east-west, four-lane road. They crossed the eastbound lanes without event, but the ten-year-old was hit when she entered the westbound lanes. The jury returned a verdict finding the cyclist sixty-three percent contributorily negligent. The Court of Appeals held that the trial court erred in determining that the cyclist was contributorily negligent as a matter of law. The issue of the ten-year-old's contributory negligence was for the jury where there was conflicting evidencing concerning whether the signal for the westbound lanes was red at the time the motorist entered the intersection and struck the cyclist in the westbound lanes such that the cyclist's negligence in crossing the intersection on the eastbound lanes on a yellow light and crossing the outside crosswalk would not have been the proximate cause of her injury. The trial court should have given the cyclist's requested instructions on driving at an excessive speed, failure to keep his car under control, failure to take evasive action, and failure to obey traffic control signals. *Savage v. Martin*, 628 N.E.2d 606 (Ill. App. Ct. 1993).

A thirteen-year-old cyclist rode through a stop sign and was hit by a motorist who had the right away. There was evidence that the cyclist's brakes malfunctioned. The jury returned a verdict in favor of the cyclist, and the trial court entered a judgment accordingly. The Court of Appeals affirmed, holding that whether the motorist exercised reasonable care under the circumstances was a question for the jury in an action for injuries suffered by the cyclist. The motorist had little or no time to honk and swerve at the same time he was attempting to apply his brakes. He was able to stop within the intersection. The jury's decision was not against the great weight of the evidence. *Cates v. Kinnard*, 626 N.E.2d 770 (Ill. App. Ct. 1994).

A six-year-old was riding her bicycle across a street and was hit by a motorist. The cyclist moved to strike the motorist's affirmative defense of comparative negligence. The trial court judge denied the motion and certified the question for interlocutory appeal. The Court of Appeals reversed, holding that Illinois' adoption of comparative negligence has had no impact on the viability of the "tender years doctrine," and the "tender years doctrine" bars the defense of comparative negligence as a matter of law in cases where a child under the age of seven is injured while riding her bike. Evidence of a violation of a traffic ordinance, therefore, is inadmissible as against a minor cyclist. *Chu v. Bowers*, 656 N.E.2d 436 (Ill. App. Ct. 1995).

A cyclist, riding on the right, was run over and killed by an overtaking semi tractor-trailer. The jury returned a verdict in favor of the cyclist's estate. The trial court judge granted the driver's motion for a new trial. The Court of Appeals held that the determination that the truck driver was liable for the cyclist's death in the collision with the truck was supported by the testimony of a witness that testified that the cyclist was riding in the street between the truck and the curb and that the cyclist appeared to lose his balance when the truck came too close. There was evidence that when the truck came to a stop after the accident without having swerved either left or right, it was between one-half and one foot from the curb. It was reasonable for the jury to infer from the foregoing evidence that the driver was driving his truck too closely to the curb, crowding the cyclist's path, and that this was the probable and proximate cause of his death. *Wojtowicz v. Cervantes*, 672 N.E.2d 357 (Ill. App. Ct. 1996).

Indiana

A thirteen-year-old cyclist, riding on a sidewalk, collided with a motorist exiting an alley. The jury returned a verdict in favor of the cyclist. The trial court judge granted a motion for a new trial. The Supreme Court reversed, holding that the statute providing that persons riding their bicycles on roadways are subject to certain provisions applicable to the drivers of vehicles was not applicable in an action for personal injuries sustained by a minor cyclist while riding his bicycle on a sidewalk. The trial court's refusal to instruct on the violation of the statute requiring bicycles to be equipped with bells or other devices capable of giving audible signals was not error in absence of evidence that the motorist listened for warnings as she approached the point of the accident. *Sheptak v. Davis*, 205 N.E.2d 548 (Ind. 1965).

A thirteen-year-old cyclist was riding against traffic and was crushed between a turning motorist and a truck. The jury returned a verdict in favor of the motorist, and judgment was entered accordingly. The Supreme Court reversed, holding that the standard of care to be applied to a thirteen-year-old cyclist who was involved in a collision with an automobile and a truck and who was charged with contributory negligence by reason of his purported violation of a safety statute forbidding driving on the wrong side of the road was that the degree of care which ordinarily would be exercised by a child of the same age, experience, intelligence, and educational level. The trial court erred in providing the instruction on the adult standard of care. *Bixenman v. Hall*, 242 N.E.2d 837 (Ind. 1968).

An eleven-year-old cyclist rode into an intersection and was hit by a motorist. The trial court judge entered judgment on a verdict in the motorist's favor. The Court of Appeals affirmed, holding that it was proper for the defense counsel to cross-examine cyclist's mother as to the extent of child's instruction, knowledge and experience with respect to riding a bicycle. Evidence tended to prove that the cyclist had received proper training. The court also held that the mere fact that there was a collision and that the cyclist was injured carried no presumption or inference of negligence on the part of the motorist and that evidence that the cyclist appeared suddenly in front of the motorist justified giving the instruction on the doctrine of sudden emergency. *LaNoux v. Hagar*, 308 N.E.2d 873 (Ind. Ct. App. 1974).

An eleven-year-old cyclist and his seven-year-old sister were riding on the same bicycle. They rode down their grandfather's driveway and stopped at the road. They saw a car approaching some 400-500 feet in the distance but because there were large dogs charging them, they decided to pull out and were hit by the motorist. The jury returned a verdict in favor of the motorist, and the cyclists appealed to the Court of Appeals, which reversed. The Supreme Court vacated, holding that the instruction that it was the duty of the driver of the motor vehicle approaching a street or private driveway intersection to use reasonable care to "discover" whether or not there is any bicycle approaching or entering such intersection and that even though the highway upon which the driver of the motor vehicle is traveling is a preferential one, there is still a duty of due care to avoid coming into collision with said bicycle, was properly refused by the trial court. If the cyclists failed to stop at the end of the driveway where they could have been seen by the motorist, the motorist could not have been negligent for failing to see them. *Thronton v. Pender*, 377 N.E.2d 613 (Ind. 1978).

A cyclist was traveling east at night with his headlights illuminated. A motorist, heading west, collided with the cyclist while making a left-hand turn. The jury returned a verdict in favor of the cyclist, and the trial court judge entered judgment accordingly. The Court of Appeals affirmed, holding that the trial court's instructions adequately presented the motorist's theories of contributory negligence to the jury notwithstanding refusal to instruct specifically on the lack of lookout and the failure to yield the right-of-way by the cyclist. *Duchane v. Johnson*, 400 N.E.2d 193 (Ind. Ct. App. 1980).

A fourteen-year-old cyclist was riding against traffic and was injured when he collided with protruding objects from a pick-up truck. The jury returned a verdict in favor of the motorist, and the trial court judge entered judgment accordingly. The Court of Appeals affirmed, holding that the trial court did not err in submitting the issue of contributory negligence to the jury where the minor cyclist was traveling in the wrong direction on the highway. The jury was properly instructed on the doctrine of the last clear chance. *Johnston v. Brown*, 468 N.E.2d 597 (Ind. Ct. App. 1984).

Iowa

A ten-year-old cyclist was riding her bicycle on the right. An overtaking vehicle hit and killed her. After a jury trial, the trial court judge entered a verdict in favor of the cyclist. The Supreme Court affirmed, holding that whether the motorist was negligent in failing to use reasonable care to pass sufficiently far enough to the left of the bicycle, which was traveling up a hill, so that the automobile struck the bicycle causing injuries which resulted in the death of a minor cyclist was for the jury. The street was thirty-feet wide, and there was no excuse suggested why the motorist could not have passed at a safer distance. *Westman v. Bingham*, 300 N.W. 525 (Iowa 1941).

A ten-year-old cyclist and a motorist entered an uncontrolled intersection at the same time with the right-of-way in favor of the cyclist. The cyclist was hit and killed by the motorist. The jury returned a verdict in favor of the motorist, and the judgment was entered accordingly. The Supreme Court held that instruction concerning the motorist's speed, lookout and right-of-way in an action against the motorist for the death of the cyclist who was struck at an intersection by the motorist was reversibly erroneous, where at no place in the instructions was the jury told that the failure to yield the right-of-way, in and of itself, whether connected with the lookout or not, would constitute negligence, and the jury was not told that the failure to keep a proper lookout, in and of itself, would constitute negligence. *Mass v. Mesic*, 142 N.W.2d 389 (Iowa 1966).

An eleven-year-old cyclist rode his bicycle onto a highway and was hit and killed by a motorist. The trial court entered judgment for the motorist after a jury verdict in her favor. The Court of Appeals reversed, holding that the evidence presented the jury with questions as to whether the cyclist was in plaint view, such that in the exercise of ordinary care, the motorist should have seen the cyclist in time to reduce the speed of her car in order to avoid hitting the cyclist and, thus, warranted the jury instruction regarding the degree of care to be observed by the motorist who observes or should observe a child riding a bicycle along a public road. *Fratzke v. Meyer*, 398 N.W.2d 200 (Iowa 1986).

A nine-year-old cyclist went through a stop sign and was hit by a motorist. After a bench trial, the court found for the motorist. The Supreme Court affirmed, holding that the evidence that the minor cyclist proceeded at a fast speed into the intersection without stopping, even though there was a stop sign, and that the motorist, who did not have a stop sign at the intersection, immediately braked her automobile and swerved, sustained the finding that the motorist was not negligent. *Davis v. Gatewood*, 228 N.W.2d 84 (Iowa 1975).

Kansas

A nine-year-old cyclist was riding with two other cyclists in single file. An overtaking truck hit and killed the nine-year-old. The Supreme Court affirmed, holding that there was evidence of negligence in the operation of the truck to effect that the truck did not deviate from its course while passing, that no warning signal was given, and that a blue thread from the cyclist's clothing found in the tread of the truck tire was sufficient for the jury. *Morrison v. Hawkeye Casualty Co.*, 212 P.2d 633 (Kan. 1949).

A six-year-old cyclist came to an intersection, stopped, and then entered the intersection where he was hit by a motorist. The jury returned a verdict in favor of the motorist, and the judgment was entered accordingly. The Supreme Court affirmed, holding that the trial court did not err in its refusal to provide the instruction on unavoidable accident. The mere fact that the accident was inevitable or unavoidable at the time of the occurrence did not entitle the motorist to the protection of the doctrine of unavoidable accident. *Gardner v. Welk*, 393 P.2d 1019 (Kan. 1964).

A motorist made a right-on-red turn while looking left. She hit and injured a cyclist. The cyclist claimed to have been riding on the sidewalk and that she was hit while in the crosswalk. The motorist claimed that the cyclist had been riding in the roadway against traffic. The jury found both parties to be fifty percent negligent, and the trial court judge entered the judgment for the motorist. The Court of Appeals affirmed. The Supreme Court reversed, holding that the use of bicycles on sidewalks was not prohibited, and if the cyclist was riding on the sidewalk, she had the right-of-way in the crosswalk. *Schallenberger v. Rudd*, 767 P.2d 841 (Kan. 1989).

Kentucky

A thirteen-year-old girl was riding on the top tube of a bicycle being operated by a fifteen-year-old boy. The thirteen-year-old was killed when they were hit by a motorist. The jury returned a verdict in favor of the motorist, and the judgment was entered accordingly. The Court of Appeals reversed, holding that the negligence of the operator of the bicycle cannot be imputed to the passenger, unless it can be shown that the passenger had an opportunity to exercise proper care for her safety and failed to do so. *Harris v. Morris*, 259 S.W.2d 469 (Ky. Ct. App. 1953).

A thirteen-year-old cyclist was riding on a roadway about two feet from the edge. An overtaking motorist hit and killed him. The jury returned a verdict in favor of the motorist, and the judgment was entered accordingly. The Court of Appeals reversed, holding that the instruction that imposed a statutory duty upon the cyclist to exercise ordinary care not to turn the bicycle from the direct course upon the highway unless it can be accomplished safely and not to turn left without signaling if any other vehicle can be affected by such movement was not justified and was prejudicial. There was no audible warning from the motorist. The cyclist's likely destination was a substantial distance ahead. *Lareua v. Trader*, 403 S.W.2d 265 (Ky. Ct. App. 1966).

A twelve-year-old cyclist was racing his bike against a friend, heading north. He came to a cross street, tried to stop, skidded, and then turned right and entered the eastbound lane where he was hit from behind by a motorist. The trial court judge directed a verdict for the motorist. The Court of Appeals reversed, holding that the evidence in the accident presented a jury question as to whether the motorist was negligent in respect to his lookout; whether he exercised ordinary care to avoid the collision; and, whether he had the last clear chance to avoid the collision. The jury should determine the cyclist's negligence, if any, applying a standard applicable to the ordinarily prudent children of the same age, education, and experience. *Williamson v. Garland*, 402 S.W.2d 80 (Ky. Ct. App. 1966).

A cyclist was riding in the outside lane of two lanes heading west. A motorist approached from the rear in the inside lane. The cyclist moved to the inside lane as the motorist was passing and was hit and killed by the motorist. The jury returned a verdict in favor of the cyclist. The trial court judge sustained a motion for judgment notwithstanding the verdict for the motorist. The Court of Appeals reversed, holding that the evidence that the cyclist and the motorist were both proceeding in the same direction and that the motorist overtook the cyclist and collided with him and that the motorist did not sound his horn or slacken his speed prior to the collision was sufficient to present questions for the jury as to the proximate cause negligence of the motorist and the contributory negligence of the cyclist. *Toombs v. Williams*, 439 S.W.2d 946 (Ky. Ct. App. 1969).

A twelve-year-old cyclist was riding at night without lights. A motorist, who had just passed another vehicle and was approaching the cyclist from the rear, ran into the cyclist, killing him. The jury returned a verdict in favor of the motorist. The Court of Appeals reversed, holding that the refusal to instruct the jury on the motorist's duty to have his headlights in working order and properly aimed was error. *Mason v. Stengell*, 441 S.W.2d 412 (Ky. 1969).

A cyclist exited a one-lane bridge and entered a one-way, one-lane section of a roadway. A motorist, approaching form the opposite direction, saw the cyclist and began to take evasive driving action. She swerved left, then right, then left before hitting the cyclist. The point of impact occurred in the cyclist's lane of travel, some ten to twelve feet out of the one-way section. The jury found for the motorist. The Court of Appeals affirmed. The Supreme Court reversed, holding that the jury instructions regarding the motorist's duty to conform the manner in which she drove to the conditions she would encounter upon reaching a one-lane roadway and bridge should have been given. The trial court's error in failing to give this instruction could not be deemed harmless. By her own testimony, the motorist indicated that she saw the cyclist while he was still on the bridge, before she reached the one-lane section, and this motorist knew, or should have known, the cyclist had the right-of-way. *Risen v. Pierce*, 807 S.W.2d 945 (Ky. 1991).

Louisiana

A ten-year-old cyclist, riding a few feet from the right-hand curb was passed by a truck traveling in the same direction. As the minor-cyclist turned slightly to his left which crashed him into the passing truck. He was knocked to the roadway and killed. The Louisiana Court of Appeals held that the truck driver, driving in close proximity to a child on a bicycle, is negligent in not expecting child to do the unexpected and dart suddenly into danger. The court also held that there was evidence at trial of excessive speed on part of truck driver to support the eleven to one vote plaintiff verdict. *Bosarge Et Ux. V. Spiess & Co. et al.*, 145 So. 21 (La. Ct. App. 1932).

Two boys, aged ten and thirteen, riding the same bicycle, entered a highway from a private lane as two automobiles approached from the west. The first driver saw the boys and stopped his auto. The second driver, insured by defendant, passed the first. He hit both boys, killing the ten-year-old. The Court of Appeals found the second driver's negligence to be the sole proximate cause of the collision between auto and bicycle. Keeping a proper lookout could have avoided the accident. The boys were not contributorily negligent. *Tate v. Norfolk and Dedham Mutual Fire Insurance Company*, 153 So. 2d 495 (La. Ct. App. 1963).

An eight-year-old cyclist was riding alone with his thirteen-year-old brother around 6:30 on a February night. The eight-year-old was hit and killed by a motorist. The judgment was entered in favor of the motorist. The Court of Appeals affirmed, holding that the failure of the motorist to see the cyclist, who was riding a bicycle, which was not equipped with a light and wearing dark clothing, was not negligence. *White v. Employers Liability Assurance Corp. Ltd.*, 210 So. 2d 580 (La. Ct. App. 1968).

A six-year-old cyclist was riding west on a straight section of a blacktop road. He was overtaken by a motorist traveling in the same direction. Immediately after the motorist returned to her lane, the young cyclist made a left-hand turn, crossing into the Eastbound lane of traffic. Defendant was traveling east at a speed of 40-50 mph, at the time, and after skidding ninety feet, collided with and injured the child. The trial court rendered judgment for defendant. The Court of Appeals affirmed the decision, holding that the defendant employed all reasonable precautions to avoid the accident and that the sudden act of the cyclist created an emergency, rendering it impossible for him to avoid the accident. *Wooten v. Wimberly*, 233 So. 2d 682 (La. Ct. App. 1970).

A ten-year-old female cyclist was "racing" at least two other minor cyclists on the city streets of Lake Charles. The finish line was approximately three bike lengths in front of an intersection. After the child passed the finish line, she looked back in the direction of her opponents. As she proceeded into the intersection, she was hit by a motorist, insured by defendant. This driver testified that he applied his brakes as soon as he saw children present in the roadway. The Court of Appeals held that the minor cyclist failed to obey yield sign at blind intersection before being struck by motorist on favored street and was contributorily negligent and, therefore, under Louisiana law, barred from recovering for personal injuries sustained in the accident. *Guillory v. Fidelity & Casualty Company of New York*, 251 So. 2d 199 (La. Ct. App. 1971).

A cyclist was riding with a friend at night, without lights. They were riding single-file when a motorist approached from the rear and hit and killed the cyclist. The trial court judge entered judgment in favor of the cyclist. The Court of Appeals affirmed, holding that the cyclist, proceeding at night on the highway without having a red reflector or warning lights visible form the rear, and who was struck and killed by an overtaking motorist, was guilty of negligence. The motorist could not be found liable under the theory of last clear chance. *L'Hoste v. Ciravola*, 261 So.2d 238 (La. Ct. App. 1972).

A seventeen-year-old cyclist was riding with three other minors. The group descended down an overpass in single-file. When the road leveled off, the boys attempted to make their way to a "neutral ground area" that separated the four-lane road consisting of two lanes of traffic in each direction. The seventeen-year-old looked back before crossing, saw the approaching vehicle, but testified that he believed he had sufficient time to clear the lane. He was hit by the automobile and injured. The trial court judge concluded that the cyclist was contributorily negligent in operation of his bicycle in that he failed to keep a proper lookout and that he improperly changed lanes. The cyclist had entered evidence of excessive speed of the motorist in the way of an accident reconstructive expert's testimony. The Court of Appeals held that skid marks left by automobile established negligence of motorist in traveling at excessive rate of speed and reversed the trial court. *Kelly v. Messina*, 318 So. 2d 74 (La. Ct. App. 1975).

A cyclist traveling the wrong way on a one-way street in New Orleans was hit by a taxicab when the driver turned onto the one-way street. The Court of Appeals held that the taxicab driver was negligent in failing to look in the direction he was traveling and that the cyclist was contributorily negligent in traveling the wrong direction on a one-way street in violation of New Orleans

City Code which applied all motor vehicle laws to bicycles and because she started across intersection after noticing taxicab was about to turn left. The court rejected the last clear chance doctrine advanced by the cyclist because it found that the taxicab driver did not have enough time to avoid accident after discovering plaintiff's peril. *Mugnier v. Check Cab Company*, 309 So. 2d 747 (La. Ct. App. 1975).

A thirteen-year-old female cyclist riding on Louisiana highway 21 was hit from behind by a motorist. The evidenced showed that the motorist was traveling in the same direction as the young girl when he noticed her in the distance. He saw another automobile coming in the opposite direction and realized the two cars would meet at approximately the same spot the cyclist was riding. He testified that he took his foot off of the accelerator and began to coast. Nevertheless, he struck and severely injured the child. The trial court found no negligence on the part of the motorist. The Court of Appeals disagreed, holding that because the motorist knew a minor cyclist was on the roadway and did not sound his horn or take other evasive action, he was negligent and that minor cyclist had not acted knowingly and recklessly so as to be contributorily negligent thereby defeating his recovery. *LaCroix v. Middle South Services, Inc.*, 345 So. 2d 136 (La. Ct. App. 1977).

A nine-year-old cyclist was riding on the right. Two motorists approached from the rear. When the cyclist noticed the first motorist, he crossed to the other side of the road and let him pass. As the second motorist was passing, the cyclist returned to the right side of the road and collided with the vehicle. The Court of Appeals affirmed, holding that a child of nine years old is capable of negligence and contributory negligence. The evidence sustained the finding that the motorist was not negligent. It was clear that the events occurred very quickly, and the motorist had little time to do more than apply the brakes. *Gladney v. Cutrer*, 440 So. 2d 938 (La. Ct. App. 1983).

An eleven-year-old cyclist was hit by an overtaking motorist. The jury returned a verdict finding the cyclist ninety-five percent negligent and the motorist five percent negligent. The trial court judge granted the cyclist's motion for judgment notwithstanding the verdict, reducing the cyclist's negligence to twenty-five percent and increasing the motorist's negligence to seventy-five percent. The Court of Appeals affirmed, holding that the trial judge's reapportionment of liability was not manifestly erroneous. The motorist testified that she had been warned that children were present on this stretch of road. The motorist startled the cyclist when she sounded her horn and continued to attempt to pass. *Robertson v. Penn*, 472 So. 2d 927 (La. Ct. App. 1985).

A cyclist was making a right-hand turn at an intersection when a City of New Orleans bus, which was overtaking him, also turned right. The cyclist and bus collided which sent the rider to roadway and under the bus. The jury found the bus driver to be seventy-five percent at-fault and the cyclist to be twenty-five percent at-fault. Under the comparative negligence statute, the jury award was apportioned accordingly. The defendant appealed. The Court of Appeals affirmed the jury finding, holding that evidence, especially the testimony of two accident reconstructionist offered by plaintiff supported the jury's apportionment. *Moliere v. Wright*, 487 So. 2d 587 (La. Ct. App. 1986).

A cyclist was riding a bicycle, heading south, against traffic without a light at around 8:30 in the evening. A motorist was backing his vehicle out of his driveway. The motorist testified that he looked in the direction of the traffic heading north, saw none, and backed his vehicle onto the road, colliding with the cyclist. The jury returned a verdict in favor of the cyclist, but found him ninety percent at fault for the accident. The Court of Appeals affirmed, holding that assigning ninety percent of the fault for the motorist/cyclist collision to the cyclist, who was riding on the wrong side of the road, with defective brakes and without the required lighting, was not manifestly erroneous. *Zachary v. Travelers Indemnity Co.*, 533 So. 2d 1300 (La. Ct. App. 1988).

A twelve-year-old cyclist rode her bike onto a street in a mostly residential neighborhood from directly in front of a parked 18-wheel truck. The defendant motorist testified that the girl appeared suddenly from behind the truck's front end and that he was unable to avoid running into her. At the bench trial, the judge apportioned eighty percent at-fault to the motorist and twenty percent to the cyclist. The defendant motorist appealed. The Court of Appeals noted that the motorist could not be found grossly negligent in trying to drive past parked truck where child darted out reduced the motorists fault to twenty percent and increased minor cyclist's fault to eighty percent. *Augustine v. Griffin*, 525 So. 2d 540 (La. Ct. App. 1988).

A cyclist riding the wrong way on a one-way two-lane road came to an intersection . Defendant motorist came to the same intersection at approximately the same time. The cyclist attempted to ride through the intersection before the motorist. The motorist looked for west-bound traffic and saw none (cyclist was riding the wrong way – eastbound) as she pulled away from a stopped position she collided with the cyclist. The trial court found that the motorist owed no duty of care to any traffic traveling the wrong-way on a one-way street. The Court of Appeals agreed after dismissing the cyclist's intersectional preemption rule. *Crump v. Ritter*, 583 So. 2d 47 (La. Ct. App. 1991).

A hearing impaired ten-year-old girl was riding her bike on a dirt road. She attempted to turn around and was struck by a motorist. At trial, the judge granted driver's motion for summary judgment. The plaintiff appealed, and the Court of Appeals reversed, stating that a high degree of care is placed on motorist who see child on or near road. The motorist has a duty to anticipate that a child, possessed of limited judgment, might be unable to appreciate impending danger. *Rhodes v. Executive Risk Consultants, Inc.*, 642 So. 2d 269 (La. Ct. App. 1994).

A cyclist was riding on the right-hand shoulder of a four-lane street. A motorist in the right lane traveling in the same direction passed the cyclist and turned right into a supermarket parking lot. The cyclist ran into the side of the automobile and was injured. At the bench trial (plaintiff had requested jury trial and appealed this issue as well) the trial judge found the motorist and cyclist seventy percent and thirty percent at-fault, respectively. The cyclist appealed. The Court of Appeals held that the trial judge had improperly applied the statute that forbids a driver from overtaking and passing other vehicles by "driving off the pavement or main traveled portion of the highway." It did not apply to bicycle being ridden on the shoulder of road. The court went on to state that although it was error to assign any fault to bicyclist, his failure to either answer motorist's appeal or appeal in his own right precluded it from increasing motorist's liability to one hundred percent. *Tenpenny v. Ringuet*, 670 So. 2d 644 (La. Ct. App. 1996).

A six-year-old cyclist riding his bicycle down a steep road that came to a T-intersection at the bottom of the hill he took a right-hand turn. He overshot his turn and collided with the defendant driver. The jury returned a defense verdict. The court affirmed the trial court judgment explaining that the evidence showed that the motorist was traveling slowly, that he reacted to the child's sudden appearance by making a sharp turn to the right and running completely off the road to avoid a collision. The evasive move was exceptional, but simply too late. The jury could reasonably conclude that the defendant driver exercised adequate care. *Jones v. Hawkins*, 708 So. 2d 749 (La. Ct. App. 1998).

A cyclist was riding east in a westbound lane that had temporarily been closed to traffic. The motorist traveling east in the eastbound lane and made a left-hand turn soon after she overtook the cyclist without using her turning signal. The motorist moved for summary judgment which trial judge denied. The motorist sought a supervisory writ that was denied by the Court of Appeals. The court found that genuine issues of material fact existed as to whether motorist was negligent in non-use of turning signal. *Schmidt v. Chevez*, 778 So. 2d 668 (La. Ct. App. 2001).

Maine

An eleven-year-old cyclist rode his bicycle down a driveway and out into a highway and was hit by a motorist. At trial, the jury found the cyclist negligent and returned a verdict for the motorist. The Supreme Judicial Court held that a bicycle is not a vehicle within the terms Maine Highway Law (in 1964) and therefore the cyclist was not required to yield the right-of-way upon entering public way from private road. Further, the court stated because cyclist stated that he looked both directions before entering a public way and did not see motorist and motorist elected not to take stand and testify to the contrary, the jury was warranted in finding motorist negligent. *Fowles v. Dakin*, 205 A.2d 169 (Me. 1964).

A fourteen-year-old cyclist was descending a hill into a "T" intersection. His hand brakes failed, and he was forced to take evasive action and veered across a highway aiming for the refuge of a shopping parking lot. A motorist passing a stopped vehicle on the right collided with the cyclist. The judge granted judgment on a directed verdict for the motorist and the cyclist appealed. The Supreme Judicial Court held that under some circumstances a motorist may pass on the right. In this case there was a factual question as to whether motorist kept a proper lookout when he passed on the right or otherwise proceeding safely. Therefore, the directed verdict was an error. There was enough evidence of motorist's negligence for jury. *Reed v. Rule*, 376 A.2d 445 (Me. 1977).

Maryland

An eleven-year-old cyclist was attempting to cross a six lane east-west highway divided by a median strip. He made it across two lanes before he was hit. At the close of the plaintiff's evidence at trial, the defendant motorist moved for a directed verdict, which was granted. The cyclist appealed. The Court of special Appeals applied the "boulevard rule" pertaining to entrance of a vehicle from non-favored street onto favored highway is not limited to its application to motor vehicles; it applies to bicycles. *Oddis v. Greene*, 273 A.2d 232 (Md. Ct. Spec. App. 1971).

A seven-year-old cyclist failed to yield the right-of-way when he rode his bicycle out of a "blind" private driveway and collided with an automobile. At trial there was a jury verdict for the minor. The Court of Special Appeals held that the "boulevard rule" applied to the facts and that the boy was negligent as a matter of law when he, as an unfavored driver, failed to yield the right-of-way. *Richards v. Goff*, 338 A.2d 80 (Md. Ct. Spec. App. 1975).

Two cyclists on a tandem bicycle were riding northbound in the southbound bus lane of coastal highway. A motorist approached the coastal highway and made a right-hand turn onto the highway. A collision between the cyclists and the motorist ensued. The motorist moved for summary judgment. The U.S. District court held that the cyclists were contributorily negligent in not riding as near to the right side of the roadway as practicable, not having a headlight and reflectors, not having a bell or other audible device on the bicycle, and riding in the wrong direction, against traffic. The cyclists did not "assume the risk" by being present on the highway on a bicycle and were not barred from recovery on that basis. The court distinguished the underlying facts from the classic boulevard case stating that the right on red statute provided invariable submission to a jury. *Longie v. Exline*, 659 F. Supp. 177 (D. Md. 1987).

Massachusetts

A fifteen-year-old cyclist was riding a bicycle along a two-way, two-lane road. As he was being passed by a truck, the cyclist fell to his left and had his left arm run over by the truck wheel. The jury found for the defendant motorist. The cyclist appealed assigning error to refusal of the trial court judge to instruct the jury that "plaintiff is not required to prove that an actual collision occurred between the truck and the cyclist before he was run over." The Court of Appeals of the First circuit held that this instruction was not necessary, explaining that the permitted instruction which allowed the jury to consider whether the distance between the moving truck and the cyclist was so close as to reflect a lack of car on the trucker's part was adequate. Kelliher v. General Transportation Services, Inc., 29 F.3d 750 (1st Cir. 1994).

Michigan

A sixteen-year-old cyclist was riding a bicycle with a seven-year-old on board. When a truck passing the boys lost control, the boys feel under the truck and were both killed. At trial, the defense moved for a directed verdict, which was granted. The Appellate Court reversed and ordered a new trial. The issue as to whether the truck passed at a safe distance should have been submitted to the jury. Whether the proximate cause of the collision between the truck's approach from the rear of the bicycle without warning, so close that the boys were startled and lost control, precluded a directed verdict. *Stockfisch v. Fox*, 267 N.W. 754 (Mich. 1936).

A cyclist was traveling downhill on a city street at 10:30 on a February night in Kalamazoo. A motorist turned onto the same street, and the two collided, killing the cyclist. The jury returned a verdict for the cyclist. The motorist appealed, citing the cyclist's contributory negligence. The Supreme Court held that in a case such as the instant one, with conflicting evidence relating to the circumstances of the accident, the speed the cyclist was traveling, and whether it constituted negligence, were issues properly put before the jury. *Swift v. Kenbeek*, 286 N.W. 658 (Mich. 1939).

A twelve-year-old cyclist was riding his bicycle down a Michigan highway shortly after dark on a December evening. An overtaking motorist passed a vehicle traveling in the opposite direction with its bright lights illuminated. Immediately after the vehicles passed one another, the motorist saw the cyclist for the first time. The motorist turned to the center of the highway, in an attempt to avoid the cyclist, but because the cyclist also veered to the left lane, the motorist collided with and killed the cyclist. After a plaintiff's verdict, the motorist appealed. The Supreme Court held that the issues of whether the cyclist was contributorily negligent in not riding on or nearer to the edge of the road and by veering to his left before the collision were proper issues for the jury. The court refused to find the cyclist contributorily negligent, as a matter of law. *Winter v. Perz*, 56 N.W.2d 276 (Mich. 1953).

After attending a party, a nineteen-year-old cyclist put his bike in the trunk of a friend's car because he felt he was too drunk to ride. They were pulled over by a police officer who administered a sobriety test to the friend and then released them. He directed the cyclist to remove his bicycle from the trunk and ride home. Approximately fifteen minutes later, an intoxicated driver hit and killed the cyclist. The defendants (the officer, the police chief and the police department) moved to dismiss. The Appellate Court held that the allegations that the officer forced the cyclist to ride his bicycle, which had no lights, at night, while the cyclist was intoxicated, sufficiently alleged a special relationship between the state and the victim so that a § 1983 action could not be dismissed based on the absence of such a relationship. Allegations were sufficient to state a claim that the officer acted with at least gross negligence and arbitrarily used his governmental power to deprive the cyclist of his life and liberty without due process. *Tittiger v. Doering*, 678 F. Supp. 177 (E.D. Mich. 1988).

Minnesota

A minor female cyclist was riding east with a companion on a two-way, two-lane highway when a male motorist, traveling west, slowed or stopped to look at her. Another motorist traveling west came upon them and veered out of the westbound lane to avoid hitting the first vehicle, thereby striking the cyclist. At trial the jury found both motorists negligent. The Minnesota Supreme Court affirmed. The cyclist was not guilty of contributory negligence. *Kellerman v. Nelson*, 122 N.W.2d 604 (Minn. 1963).

A nine-year-old cyclist was waiting at a stoplight next to a truck. When the light turned green, the boy started to ride. The truck made a right-hand turn and ran over the boy. The trial judge granted the motorist's motion for a directed verdict at the close of the cyclist's case. The Minnesota Supreme Court held that the issues of the functioning of the trucks rear signal light and driver's "look out" raised jury issues precluding a directed verdict. *Peckskamp v. McDowall*, 165 N.W. 2d 254 (Minn. 1969).

A cyclist was between a stopped car and the curb waiting for a red light to change. When the light turned green, the motorist turned right and collided with the cyclist. The jury found for the cyclist. The motorist moved for judgment notwithstanding the verdict, which the trial judge granted. The Supreme Court of Minnesota held that this intersectional collision occurred when the cyclist created the dangerous situation in that he "was not legally entitled to position his bicycle between defendant's car and the curb;" that he should have approached from the rear where he could see the motorist's turning signal and been clear of the motorist's "blind spot." *Sikes v. Garrett*, 262 N.W.2d 681 (Minn. 1977).

A six-year-old cyclist was riding with a friend on a city street when he lost control of his bike. He somersaulted over the handlebars. His mouth caught the rear turnbuckle hook of a legally parked truck's homemade camper. The trail judge granted a directed verdict, and the cyclist appealed. The Court of Appeals held that the case should have gone to the jury. The question of the duty of care, while usually a legal issue for the court in close cases, is an issue of foreseeability for the jury. *Oswald v. Law*, 445 N.W.2d 840 (Minn. 1980).

A cyclist was riding on a sidewalk at night with his bicycle headlight illuminated. A motorist, driving a truck, pulled out of a "blind alley" and crossed the sidewalk. The cyclist ran into the front of the truck and was injured. After a bench trial, the motorist was found to be free of negligence. The Court of Appeals held that Minnesota Statute §169.31 applied and determined that the motorist was negligent as a matter of law and also attributed negligence on the cyclist's part before remanding the case for reconsideration by the trial court. *Mortensen v. Lucky 7 Gambling, Inc.*, 1993 WL 500526 (Minn. Ct. App.).

Mississippi

A nine-year-old cyclist and his older brother were riding down a steep hill. A truck overtaking them collided with the nine-year-old. The truck driver argued that the young cyclist had turned into the side of his truck. The boy maintained that he had traveled in a straight line. The jury found for the cyclist, and the motorist appealed. The Supreme Court held that the motorist could have turned away from the cyclist as he had approximately eighteen feet within which to pass to the left of the cyclist. The motorist did not sound his horn. He was passing children on the roadway and should have anticipated unusual and impulsive actions. *McMinn v. Lilly*, 60 So. 2d 603 (Miss. 1952).

Two minor cyclists, riding in single file, were overtaken by a motorist. The motorist struck the second cyclist in line as he attempted to pass. The trial court judge granted the motorist's directed verdict motion. The Supreme Court reversed, holding that issues of whether the motorist failed to keep a proper lookout and to use reasonable care in overtaking and passing the cyclist were questions for a jury. *Moak v. Black*, 92 So. 2d 845 (Miss. 1957).

Missouri

A fifteen-year-old cyclist was riding down a city street when a city bus overtook him. The cyclist was killed when he and the bus collided. The cyclist was passing a parked beer truck just as the bus was overtaking him. The cyclist was killed when he and the bus collided. The city bus company and the brewery were defendants at trial. The jury found for the defendants. The Supreme Court reversed, holding that the violation of the city ordinance concerning parking as near to the right as is practical were admissible in evidence. *Floyd v. St. Louis Public Service Company*, 280 S.W.2d 74 (Mo. 1955).

A thirteen-year-old cyclist was riding at night, without a light on his bicycle. A motorist, traveling in the opposite direction, passed a slower motorist, thereby entering the cyclist lane. The motorist and the cyclist collided, and the cyclist was killed. The jury returned a verdict in favor of the motorist. The Supreme Court affirmed, holding that the operation of a bicycle at night without lights constituted negligence and caused or directly contributed to the accident. *Burt v. Becker*, 497 S.W.2d 411 (Mo. 1973).

A fifteen-year-old cyclist was riding in the right lane. A motorist approached from the rear, sounded his horn as he was passing, and the cyclist turned to the left and was hit by the motorist. The jury returned a verdict in favor of the motorist. The trial court granted a new trial. The Court of Appeals affirmed, holding that the evidence was sufficient to permit the jury to find that the motorist's audible warning of his intent to overtake and pass the cyclist, who, according to his evidence, was startled into pulling left in the path of the overtaking motorist when the horn was sounded was not guilty of contributory negligence as a matter of law. *Hubbard v. Lathrop*, 545 S.W.2d 361 (Mo. Ct. App. 1977).

A northbound cyclist ran into the side of a southbound motorist who had made a left-hand turn directly in front of the cyclist. At trial, the cyclist was assigned sixty percent fault for his failure to keep a careful lookout. The Court of Appeals held that the evidence was insufficient to warrant an instruction on the comparative negligence of the cyclist without evidence that the cyclist had the means and ability to avoid the collision if he had kept a careful lookout. *Dick v. Carbon*, 926 S.W.2d 172 (Mo. Ct. App. 1996).

Montana

An eight-year-old cyclist was riding down an alley. At the end of the alley he intended to make a right-hand turn and head north; however, he was traveling too fast to maintain control of his bicycle and ended up in the southbound lane where he collided with a motorist. The trial court entered judgment for the cyclist apportioning fifty percent liability to both parties. The motorist appealed. The Supreme Court affirmed the apportionment, and there was sufficient evidence to support the jury's finding that the motorist was negligent by failing to keep a proper lookout and have his vehicle under sufficient control where he had up to three seconds to see the cyclist and avoid the accident. The plaintiff's accident reconstruction expert testified that the motorist could have completely stopped his vehicle in two to three seconds. *Okland v. Wolf*, 850 P.2d 302 (Mont. 1993).

Nebraska

A fourteen-year-old boy, one of three minor cyclists, was riding down a highway in single file, in second position. An overtaking motorist hit and killed him. The trial court entered judgment for the cyclist. The motorist appealed. The Supreme Court held that it was reversible error for the trial court not to instruct on the duty of the cyclist to maintain a proper lookout and control of his vehicle (bicycle). There was evidence, in the motorist's testimony, that the cyclist had violated these duties. *Sacca v. Marshall*, 146 N.W.2d 375 (Neb. 1966).

An eleven-year old cyclist approached an intersection. She yielded to a gravel truck. As she began to pedal, she saw a motorist was going to run into her and could not avoid the collision. At the close of the cyclist's case in chief, the judge dismissed the case, and the cyclist appealed. The Supreme Court held that, under the circumstances of the case, the evidence established, as a matter of law, that the cyclist's negligence was not the sole proximate cause of the accident. The case should go to the jury. *Gadeken v. Langhorst*, 226 N.W.2d 632 (Neb. 1975).

A fourteen-year-old cyclist was riding alongside a friend at night. The fourteen-year-old was hit by an overtaking motorist and killed. The jury returned a verdict in favor of the cyclist's estate. The Supreme Court affirmed, holding that evidence of the cyclist's contributory negligence was properly admitted. There was ample evidence from which the jury could have determined that the defendant was negligent with regard to speed, lookout, and proper control. *Caradori v. Fitch*, 263 N.W.2d 649 (Neb. 1978).

A fifteen-year-old cyclist, riding down a sidewalk, came upon a pickup truck stopped across the sidewalk. The cyclist looped around the back and according to the witnesses in the pickup truck, out into the roadway without stopping or looking for traffic. He collided with a motorist's vehicle. The trial judge entered summary judgment for the motorist. The Supreme Court held that the cyclist's failure to look for approaching traffic before entering the street was the proximate cause of the collision between the motorist and the cyclist, precluding the cyclist's right to recover. The situation was created so quickly that the motorist had no opportunity to avoid the accident and is, therefore, not liable. *McFarland v. King*, 341 N.W.2d 920 (Neb. 1983).

A fourteen-year-old was one of three minor cyclists riding down a sidewalk. While the fourteen-year-old was in a crosswalk, he was hit by a motorist who was making a right-hand turn on a green light. The trial judge granted the motorist's motion for directed verdict, based on the cyclist's negligence in failing to keep a lookout and yield the right-of-way to all vehicles upon entering the highway. The Court of Appeals reversed, holding that the cyclist was legally in the crosswalk and was entitled to the same rights as a pedestrian. *Luellman v. Ambroz*, 516 N.W.2d 627 (Neb. Ct. App. 1994).

A cyclist, riding in a crosswalk, was hit by a motorist making a right-hand turn from a stopped position. The jury returned a verdict, finding the cyclist fifty percent negligent and the motorist fifty percent negligent, resulting in a verdict for the motorist under Nebraska's comparative negligence statute. The Court of Appeals refused to set aside the jury verdict. There was conflicting evidence as to who entered the intersection first. There was sufficient evidence to support the instruction given to the jury on the cyclist's failure to yield the right-of-way. *Radecki v. Burns*, 2001 WL 682050 (Neb. Ct. App.).

Nevada

A cyclist was straddling his bicycle while walking in a crosswalk when he was hit by a motorist who had a yield sign. The trial judge entered judgment in favor of the motorist. The Supreme Court held that the cyclist was a pedestrian, despite the fact that he was straddling his bike at the time he was hit. He was entitled to a jury instruction on negligence per se. *Del Piero v. Phillips*, 769 P.2d 53 (Nev. 1989).

New Hampshire

An eleven-year-old boy was riding his bike through an air force base. He was overtaken by a tractor-trailer. The boy lost control of his bike, just as the trucker was passing him, when his eyeglasses fell off his face, and he grabbed them. He veered left and was run over by the rear wheels of the trailer with fatal results. The trial court, in a bench decision, entered judgment for the motorist. The First Circuit Court of Appeals affirmed the trial court's holding that the evidence supported the conclusion that the motorist's conduct was not a proximate cause of the accident. The District Court's decision to credit the motorist's expert witness over the cyclist's expert witness was not clearly erroneous. *Fernberg v. T.F. Boyle Transportation, Inc.*, 889 F.2d 1205 (1st Cir. 1989).

A thirteen-year-old cyclist was riding with two companions. All were on bikes. The cyclists were riding in a westerly direction on the left side of the road. The first two cyclists crossed to the right side of the road. When the thirteen-year-old plaintiff attempted to cross, he was hit by a following motorist. Judgment was entered on a jury verdict in favor of the motorist. The Supreme Court held that where the evidence showed the motorist let off the gas upon seeing the boys, jammed on the brakes when the cyclist turned in front of him and where the cyclist chose to continue to cross the roadway in spite of the approaching motorists, the jury could reasonably have concluded that the motorist was not negligent in the operation of his vehicle. The opinion testimony of the responding officer was properly excluded because he was not an accident reconstruction expert, and his opinion concerning fault and the manner and cause of the collision would have involved a mixed question of law and would not have assisted the jury. *Johnston v. Lynch*, 574 A.2d 934 (N.H. 1990).

New Jersey

A sixteen-year-old cyclist was riding down a city street with cars parked along the curb. As he passed the defendant's vehicle, the front door opened. The cyclist veered to his left, but ran into the door falling to the roadway. A following motorist collided with the cyclist. At trial, the motorist, who was overtaking the cyclist, was dismissed from the suit. The Superior Court held that there was evidence at trial that could have had a jury find negligence on the motorist's part. The motorist traveled sixty-two feet or more after hitting the cyclist, supporting the inference that the motorist was traveling at an excessive speed. *Volksen v. Kelly*, 79 A.2d 319 (N.J. Super. Ct. App. Div. 1951).

A cyclist, riding at night on a bicycle equipped with a headlight and rear reflectors, was hit by an overtaking motorist. The trial court judge granted the motorist's motion to dismiss at the close of the cyclist's case in chief. The Appellate Court reversed, holding that the evidence presented a question of negligence and contributory negligence for the jury. The jury might properly have found that the cyclist had reason to rely on the motorist seeing him. The area was well lit, and other disinterested witnesses testified that the cyclist was visible. *Duffy v. Cratsley*, 102 A.2d 63 (N.J. Super. Ct. App. Div. 1953).

An eleven-year-old cyclist was run over an intersection by a motorist driving a van. There was a dispute as to whether the cyclist was riding on the street or the adjoining sidewalk. The Superior Court held that the motorist requested instructions which spoke in terms of a "driver of a vehicle" or "highway" were not to be presented to the jury absent proof by a fair preponderance of credible evidence that the cyclist was riding on the street. The burden of this proof is on the motorist. *Gibson v. Arrowhead Conditioning Company, Inc.*, 602 A2d 800 (N.J. Super. Ct. Law Div. 1991).

New Mexico

A seven-year-old cyclist entered an intersection and was hit and killed by a speeding motorist. After a bench trial, the judgment was entered in favor of the cyclist's estate. The Court of Appeals held that even if the cyclist was riding on the wrong side of the road and entered the intersection without stopping, recovery for his death due to the injuries he sustained in a collision with a motorist are not precluded. The motorist was greatly exceeding the speed limit

(80 miles per hour in a residential area), and the accident could not have been avoided by the cyclist. *Wilson v. Wylie*, 518 P.2d 1213 (N. M. Ct. App. 1974).

A fourteen-year-old cyclist riding a unicycle was hit by a motorist. The trial judge granted the motorist's motion for summary judgment. The court of Appeals held that the material issues of fact precluded the rendition of summary judgment. The court determined question of whether the cyclist was riding or walking at the time he was struck and whether he was traveling against traffic were for the fact finder. *Wilkinson v. Randel*, 545 P.2d 95 (N.M. Ct. App. 1975).

A cyclist, riding at night, equipped with a floodlight and reflectors on his pedals and bicycle seat, was hit from behind by a motorist. The trial court entered a judgment in favor of the cyclist. The Court of Appeals affirmed, holding that it was uncontroverted that the motorist failed to keep a proper lookout, admitted to being "tipsy" after consuming beer, and this negligence was the proximate cause of the accident. *Matney v. Evans*, 598 P.2d 644 (N.M. Ct. App. 1979).

New York

A fifteen-year-old cyclist was straddling his bike, waiting for a red light to turn green. A tractor-trailer next to him made a right-hand turn, causing the trailer to strike and injure the cyclist. The trial judge entered a judgment of "no cause for action" against the cyclist. The Appellate Division reversed, holding that the motorist's turn was a violation of vehicle and traffic law because of the unsafe turn and that the cyclist was not contributorily negligent. The cyclist was straddling his bike on the gravel shoulder of the highway where he had a legal right to be. *Ziparo v. Hartwells Garage*, 75 A.D.2d 997 (N.Y. App. Div. 1980).

A cyclist was riding through Central Park when an overtaking city bus hit him. The transit authority appealed from a judgment entered in favor of the cyclist. The Appellate Court held that the circumstantial evidence supporting a finding of liability on the part of the transit authority, including accident reconstruction expert testimony. Witness testimony that the bleeding cyclist's declaration that he had been hit by the bus was properly admitted as a spontaneous declaration and was, therefore, admissible as an exception to the hearsay rule. *Flynn v.*

Manhattan and Bronx Surface Transit Operating Authority, 94 A.D.2d 617 (N.Y. App. Div. 1983).

A cyclist was riding at night, without lights or reflectors. An approaching motorist struck and killed him as the motorist overtook another motorist. The trial court judge dismissed the case at the end of the plaintiff's case. The Court of Appeals affirmed, holding that there was no evidence of negligence on the part of the motorist. He waited for the opposite lane to clear before passing. Witnesses agreed that there was no time to avoid the collision. The cyclist was negligent in wearing dark clothing and not having his bicycle equipped with lights or reflectors. *Weise v. Lazore*, 99 A.D.2d 919 (N.Y. App. Div. 1984).

A thirteen-year-old cyclist was riding on the shoulder and was attempting to cross the road at forty-five degree angle. The motorist was unable to avoid hitting and killing her. A judgment in favor of the motorist was entered. The Appellate Division reversed the holding that the trial court's failure to give a "proper lookout" charge was reversible error. The trial court's recital of the general applicable to all negligence cases that a person must exercise the degree of care that an ordinary, careful and prudent person would exercise under like circumstances was inadequate. *Pedersen v. Balzan*, 117 A.D.2d 933 (N.Y. App. Div. 1986).

A cyclist died as a result of a head-on collision with a motorcyclist who had maneuvered his motorcycle around a stopped van. The van's driver and the motorcyclist were both defendants. The trial court entered a judgment in favor of both defendants. The Supreme Court, Appellate Division, affirmed, holding that the trial court's refusal to allow police officer be called on rebuttal for the purposes of impeachment as to whether the van was double-parked was not abuse of discretion. The instruction on the violation of a section of the Vehicle and Traffic Law governing operation of bicycles was properly held. *Frias v. Fanning*, 119 A.D.2d 796 (N.Y. App. Div. 1986).

A cyclist was injured when a left-turning motorist crossed in front of him. After the jury returned a verdict in favor of the motorist, the trial court set it aside as being against the great weight of the evidence. The Supreme Court, Appellate Division, held that a left-turning motorist was negligent in either failing to see the cyclist or in trying to cross in front of him when it was hazardous to do so. *Hernandez v. Joseph*, 143 A.D.2d 632 (N.Y. App. Div. 1988).

A northbound cyclist collided with a southbound left-turning motorist. The trial judge granted the cyclist's motion for judgment as a matter of law. The Supreme Court, Appellate Division, held that the motorist was negligent as a matter of law. The motorist admitted that he did not see the cyclist approaching from the opposite direction. If he did see the cyclist, but disregarded him, he was negligent in attempting to cross in front of him when it was hazardous to do so. *Burns v. Mastroianni*, 173 A.D.2d 754 (N.Y. App. Div. 1991).

A cyclist was hit and injured by a motorist as she rode across an avenue in New York City. A jury held in favor of the cyclist. The Supreme Court, Appellate Division, affirmed, holding that no reasonable view of the evidence could support the conclusion that the lack of a bell on the bicycle was the proximate cause of the accident and, therefore, there was no error in declining to instruct the jury on the statutory violation. The evidence showed that the cyclist never saw the motorist approaching and, therefore, the lack of the bell was irrelevant. *Cranston v. Oxford Resources Corp.*, 173 A.D.2d 757 (N.Y. App. Div. 1991).

A minor cyclist was hit by a motorist after the cyclist pulled out form behind a UPS delivery truck. The motorist and UPS were the defendants at trial. The jury returned a verdict finding UPS one hundred percent liable. The trial judge granted UPS' motion to set aside the verdict. The Supreme Court, Appellate Division, held that the delivery truck was not negligent as a matter of law. The truck was lawfully parked, excepted from the "no parking" ordinance because the driver was stopped "temporarily for purposes of and while actually engaged in loading or unloading merchandise or passengers." *Reid v. Lichinchi*, 215 A.D.2d 639 (N.Y. App. Div. 1995).

A cyclist was "cut off" by a motorist and as a consequence, crossed the centerline and collided with another motorist. The trial judge dismissed the second motorist who had been brought into the suit in a third-party action. The Supreme Court, Appellate Division, affirmed the trial court, holding that the second motorist had established a complete defense to the third-party action where the evidence showed the head-on collision was caused solely by the cyclist crossing into the second motorist's lane. There was no evidence of negligence on the second motorist part. The first motorist was not entitled to summary judgment as material issues of fact existed as to whether the first motorist negligently caused the accident. *Ryan v. Ratchuk*, 221 A.D.2d 1021 (N.Y. App. Div. 1995).

A cyclist was overtaken by a motorist who then made a right turn. The cyclist collided with the motorist's vehicle. The jury returned a verdict in favor of the motorist, and the trial judge granted the cyclist's motion to set aside the verdict. The Supreme Court, Appellate Division, reinstated the verdict, holding that the evidence supported the jury's finding. The motorist was free from negligence where the motorist had his right directional signal on for 150 fee before the collision and was traveling between five and ten miles per hour. *Kash v. Kroeger*, 222 A.D.2d 1101 (N.Y. App. Div. 1995).

North Carolina

A fourteen-year-old cyclist was a passenger on a bicycle being operated by a friend. They were riding against traffic when they were hit by a motorist. A motion for a judgment of non-suit was granted by the trial court judge. The Supreme Court reversed, holding that the passenger was a guest without power to control his bicycle and the negligence of the operator could not be imputed to him. *Hensley v. Briggs*, 52 S.E.2d 5 (N.C. 1949).

A fifteen-year-old boy was riding on a downhill section of a highway when an overtaking passenger bus hit and killed him. The bus driver waited for approaching traffic to clear and then gunned his engine to pass. At the close of plaintiff's case, the trial judge granted the bus driver's motion for nonsuit. The Supreme Court reversed, holding that there was evidence of actionable negligence where the bus driver should have realized the cyclist was oblivious to the bus' presence, and the unexpected and startling noise of the accelerating bus engine was sufficient to take the case to the jury. *Webb v. Felton*, 147 S.E.2d 219 (N.C. 1966).

A thirteen-year-old cyclist was hit and killed by a following motorist. At trial, the motorist's motion for nonsuit was allowed. The Supreme Court held that the mere fact of a collision of the motorist with the cyclist ahead, although not conclusive, furnished some evidence that the motorist was negligent as to speed, following too closely, or in failing to keep a proper lookout. Questions of the motorist's negligence, whether this negligence was the proximate cause of death and whether the cyclist was contributorily negligent are jury questions. *Champion v. Waller*, 150 S.E.2d 783 (N.C. 1966).

A cyclist was riding down a highway and attempted a left-hand turn into his driveway. He was hit from the rear by a following motorist. The trial judge allowed the motorist's motion of nonsuit. The Supreme Court affirmed, holding that the cyclist was not entitled to recover, under North Carolina's contributory negligence laws, when he turned his bike to the left and into the overtaking motorist's path without looking back or giving an other warning that he was going to turn left. *Lowe v. Futrell*, 157 S.E.2d 92 (N.C. 1967).

A five-year-old boy was learning to ride a bike. Her teenage uncle was walking behind her as a motorist traveling in the opposite direction approached. The young girl began to veer to the left and into the motorist's side of the roadway. The trial judge granted the motorist's motion for compulsory nonsuit. The Court of Appeals reversed, holding that the evidence presented questions for the jury as to whether the motorist, who observed the child cyclist when she was approaching, 1000 feet away, was negligent in driving her vehicle at a greater speed than was reasonable and prudent. A legal duty rested upon the motorist to exercise due care to avoid injuring the children. The very presence of a five-year-old, riding a bicycle on the highway is, in itself, a danger signal to an approaching motorist. *Boyd v. Blake*, 159 S.E.2d 256 (N.C. Ct. App. 1968).

A thirteen-year-old cyclist and his cousin were riding south on a highway. Two motorists were following behind them. The first motorist passed both of the cyclists without incident. The second motorist hit the second cyclist in line, the thirteen-year-old plaintiff. The trial judge allowed the defendant's motion for directed verdict at the close of the cyclist's evidence. The Court of Appeals reversed, holding that because the motorist failed to sound his horn, attempted to pass at an intersection, failed to drive on the right side of the road, failed to keep a vigilant lookout, and failed to keep his vehicle under control, his negligence was for the jury to determine. *Bell v. Brueggemyer*, 242 S.E.2d 392 (N.C. Ct. App. 1978).

An eleven-year-old cyclist exited his driveway and as he crossed the centerline, he was hit by a motorist. The trial court judge granted a directed verdict for the motorist. The Court of Appeals reversed, holding that the issue of whether the motorist could have seen the minor cyclist exiting the driveway or entering highway and was, thus, under a duty to proceed with caution were for the jury. *Adkins v. Carter*, 252 S.E.2d 268 (N.C. Ct. App. 1979).

An eleven-year-old cyclist, with his seven-year-old brother on board his bicycle, rode out of a driveway and onto the road. The eleven-year-old saw an approaching motorist, but testified that he thought he had enough time to turn around and return to the driveway. The boys were hit by the motorist. The trial court judge granted the motorist's motion for directed verdict at the close of the cyclist's case. The Court of Appeals reversed, holding that the evidence presented a jury question as to whether the motorist, in exercising a proper lookout, could have seen the cyclists from 200 feet away as he approached the driveway they were exiting from and whether, by maintaining a proper lookout and exercising due care and caution thereafter, he could have avoided the collision. *Wallace v. Evans*, 208 S.E.2d 193 (N.C. Ct. App. 1982).

A fifteen-year-old cyclist rode his bicycle down his family's gravel driveway and onto the roadway. He heard a motorist approaching, and even recognized the distinct sound as one of his neighbor's vehicles. Upon entering the roadway, the cyclist crossed the centerline twice in the process of completing a circle as was his and other children's custom in the neighborhood, and was hit by the motorist's vehicle. At trial, the trial court judge granted the motorist's motion for a directed verdict. The Court of Appeals held that the boy was contributorily negligent in failing to exercise the ordinary care of a reasonably prudent person under the circumstances, and the cyclist's failure was the proximate and contributing cause of the accident. The cyclist had ridden across the highway and attempted to cross back over the centerline without looking back. There was sufficient evidence to justify the submission of the cyclist's negligence claim against the motorist to the jury on the issue of last clear chance. The motorist had, on prior occasions, encountered the cyclist's habit of moving into the left lane as to allow the motorist to pass on the right. The motorist should have been on notice that the cyclist may continue into the left lane and was under a duty to attempt ascertain the cyclist's next move before passing. *Lewis v. Brumbles*, 349 S.E.2d 323 (N.C. Ct. App. 1986).

North Dakota

A nine-year-old cyclist rode his bicycle into an intersection and was hit by a motorist. The judge, sitting as a trier of fact, entered judgment for the cyclist. The Supreme Court affirmed, holding that the cyclist looked both ways before entering the intersection, and the rule giving the vehicle on the right the right-of-way was inapplicable as evidence showed that the cyclist had entered the intersection at a distinct and appreciable interval before the motorist, thus giving the cyclist the right-of-way. *Kleinjan v. Knutson*, 207 N.W.2d 247 (N.D. 1973).

A fourteen-year-old cyclist was riding on the handlebars of a friend's bicycle. They were "sideswiped" by a motorist. The jury returned a verdict in favor of the motorist, and the judgment was entered accordingly. The Supreme Court reversed, holding that the negligence of the bike operator could not be imputed to the passenger on a theory of joint enterprise or joint venture and, therefore, the passenger was not guilty of contributory negligence. *Dimond v. Kling*, 221 N.W.2d 86 (N.D. 1974).

Ohio

A minor cyclist was riding at night on the shoulder of a road. He had a headlight and rear reflector. He was hit by a motorist. The jury returned a verdict in favor of the cyclist, and the judgment was entered accordingly. The Court of Appeals held that bicycles are vehicles within the meaning of that terms as used in the statutes pertaining to vehicles upon the highways. Therefore, a bicycle operated upon a street or highway between one-half hour after sunset to one-half hour before sunrise must display a light exhibiting a white light visible for a distance of 500 feet to the front of a bicycle and a red light visible for a distance of 500 feet to the rear of a bicycle. It was a jury question as to whether or not the negligence per se of the cyclist was the proximate cause of the collision. *Morris v. Stone*, 292 N.E.2d 891 (Ohio Ct. App. 1972).

A northbound cyclist came upon a stalled van in her lane at an intersection. They had the green light. The cyclist passed the stalled van on the right, in the same lane the van occupied. A southbound motorist attempting a left-hand turn through the intersection turned in front of the cyclist after being waived through by the van driver. The cyclist collided with the passenger side of the motorist's vehicle. The trial court judge entered judgment for the motorist. The Court of appeals reversed, holding that the motorist was negligent in making a left-hand turn in front of the cyclist, who had the right-of-way. The fact that the motorist looked and did not see the cyclist, who was in plain sight, did not absolve the motorist of negligence. *Falck v. Proos*, 455 N.E.2d 1020 (Ohio Ct. App. 1982).

An eight-year-old cyclist, riding with an eleven-year-old cyclist, was hit and killed by an overtaking motorist. The trial court judge set aside a motorist verdict. The Court of appeals affirmed, concluding that evidence at trial was such that "the trial court could reasonably reach the conclusion that it did on the motion for a new trial and that the court did not abuse its discretion in granting the motion." Evidence at trial showed that the motorist saw the boys and had ample time to obey assured clear distance statute. There was no oncoming traffic, and the motorist did not warn the cyclist of his presence by sounding his horn. *Sanderson v. Jewell*, 1986 WL 5902 (Ohio Ct. App.).

A ten-year-old cyclist was crossing a two-lane road at the top of a hill. The cyclist was about a quarter of the way across the road when he saw an approaching motorcyclist. The cyclist sped up to clear the road, the motorcyclist veered left and collided with the cyclist. The motorcyclist sustained serious head injuries and brought suit against the cyclist. The trial court judge entered judgment in favor of the defendant cyclist. The Court of Appeals affirmed, holding that the motorcyclist's requested instruction on assured clear distance rule was obviated by the jury's finding that the cyclist was not negligent. *Geul v. Robinson*, 1986 WL 14470 (Ohio Ct. App.).

A cyclist, in violation of Ohio statute, was riding against traffic on the shoulder. She was carrying a box in one hand and looking down while she rode. A motorist tailgating behind a larger vehicle, did not see the cyclist, and they collided. The jury returned a verdict finding the cyclist sixty percent negligent, and the motorist forty percent negligent. The Court of Appeals affirmed stating that the jury verdict was supported by competent, credible evidence, and the judgment of the trial court was not against the manifest weight of evidence. *Rachow v. Shutters*, 1987 WL 5887 (Ohio Ct. App.).

A cyclist, riding as far to the right as was practical, was hit from the rear by a motorist who had stated that he "lost her in the sunlight." The jury returned a verdict finding the cyclist twenty percent negligent. The Court of Appeals held that the jury's finding that cyclist was twenty percent contributorily negligent was based upon a violation of a duty that did not apply in circumstances of this case. A cyclist is not required to continually look to the rear to ascertain whether a motorist approaching from the rear is going to strike. *Howard v. McKitrick*, 1987 WL 13837 (Ohio Ct. App.).

A cyclist was riding behind a police cruiser. The policeman received an urgent radio call and turned into a private drive to turn around. The cyclist collided with the cruiser and was injured. The trial court entered judgment in favor of the policeman. The Court of Appeals held that the witness, a police officer, was not qualified to offer expert or lay opinion as to the proximate cause of the cyclist's collision with the police cruiser. The officer did not witness the accident and was not an accident reconstruction expert with experience and knowledge to observe the scene and add scientific, technical or specialized knowledge to the evidence. The officer's testimony was a clear invasion of the jury's province on the precise ultimate fact and issue. *Hatfield v. Andermatt*, 561 N.E.2d 1023 (Ohio Ct. App. 1988).

A seven-year-old cyclist was visiting a friend. The friend's mother allowed the boys to ride in the street. A motorist hit the seven-year-old. The jury returned a verdict finding the cyclist thirty-seven percent negligent, the motorist thirty-seven percent negligent, and the cyclist's parents twenty-six percent negligent. The Court of Appeals affirmed, holding that the evidence made a question for the jury as to whether cyclist's parents negligently instructed the cyclist in bike safety and as to whether the bicycle was properly equipped with lights and reflectors. *Darwish v. Harmon*, 633 N.E.2d 546 (Ohio Ct. App. 1992).

A nine-year-old cyclist was riding against traffic with a friend. As an oncoming motorist approached, the friend crossed in her path without incident. The nine-year-old attempted to cross and was hit by the motorist. The trial court judge granted motorist's motion for directed verdict. The Court of Appeals reversed, holding that although the motorist had been driving within the speed limit, and had braked upon observing the two girls approaching, there was evidence that she had been inattentive as to the cyclist. The motorist testified that she assumed the second cyclist would not follow the first. There was evidence that children riding on this section was commonplace, putting the motorist on evidence. *Gaffney v. Sexton*, 1993 WL 39609 (Ohio Ct. App.).

A thirteen-year-old cyclist was overtaken by a school bus that collided with him. There were several witnesses offering as many versions. The jury found the motorist fifty-one percent negligent and the cyclist forty-nine percent negligent. The Court of Appeals held that the trial court should have instructed an assured-clear-distances (cyclist was discernable and in bus driver's directional line of travel), on duty to continue to look (motorist's testimony that she did not have cyclist in her sight at time of impact) and erred on instruction regarding meaning of negligence per se (instruction suggested that failure to comply with rules or regulations at issue could constitute negligence rather than negligence per se were confusing and thus may have resulted in an improper verdict). *Siders v. Reynoldsburg School District*, 650 N.E.2d 150 (Ohio Ct. App. 1994).

Oklahoma

A ten-year-old cyclist was riding eastbound on the right side of a highway. He crossed into the westbound lane and was hit by a westbound motorist. The trial court judge entered judgment in favor of the cyclist. The Supreme Court affirmed, holding that whether the motorist was negligent and whether such negligence was the proximate cause of collision was properly put before the jury. It was for the jury to determine whether it was the motorist's duty, under all the circumstances and seeing the young cyclist riding in the other lane, to have sounded automobile's horn, and/or to have brought the auto under such control that the collision would not have occurred, and whether his failure proximately caused the collision. *Bready v. Tipton*, 407 P.2d 194 (Okla. 1965).

Oregon

A cyclist was riding on the Pacific Highway while delivering newspapers for his son at 6:00 on a January morning. The sunrise was at 7:20 on that day. An overtaking truck did not see him and hit and killed him. The cyclist had a headlamp and rear reflector. The trial court judge entered judgment for the motorist after the jury verdict. The Supreme Court reversed, holding that it was error to submit specifications of cyclist's negligence and that failure to give plaintiff's requested instruction, constituted error. *Spence v. Rasmussen*, 226 P.2d 819 (Or. 1951).

A fourteen-year-old cyclist came to a T-intersection. He stopped and waited for traffic to clear. He entered the intersection and was hit by a motorist. The trial court judge rendered judgment pursuant to an order of involuntary nonsuit. The Supreme Court reversed, holding that the posted speed limit is only presumptively lawful. Its lawfulness may be rebutted by a showing of circumstances indicating that a reasonably prudent person would be traveling at a legal speed. Whether the motorist who was proceeding on the road posted for 25 m.p.h. maximum speed limit, whose view of traffic approaching from the south was obstructed by vegetation, whose vehicle had collided with northbound right-turning cyclist after his automobile had skidded approximately forty-eight feet and whose vehicle had traveled approximately feet beyond of sixty foot two inches of skid marks on surface that was covered by loose gravel before coming to a stop was a question for the jury. *Loofburrow v. Flynn*, 496 P.2d 220 (Or. 1972).

A cyclist was riding down a sidewalk with her husband in Corvallis. A motorist, exiting from a parking lot, "nosed" out into the sidewalk, colliding with the cyclist. The trial court entered judgment of a special verdict finding the motorist, free from negligence. The Court of Appeals reversed, holding that the motorist was negligent in failing to stop his automobile before entering the sidewalk as he was leaving the parking lot. The fact that the cyclist was riding on a sidewalk in the business district in violation of its ordinance did not constitute negligence per se. The ordinance was intended to protect pedestrians from injury from cyclists. *Reynolds v. Tyler*, 670 P.2d 223 (Or. Ct. App. 1983).

A cyclist was riding westbound on 19th Avenue in Eugene. He wanted to turn left, but because of a steady stream of eastbound traffic, he was forced to stop in traffic and wait for an opening. A motorist, traveling in the same direction, hit him from behind. The trial court judge entered judgment for the motorist. The Court of Appeals reversed, holding that the jury instruction concerning the cyclist's duty to keep to the right side of the roadway while traveling with traffic had no application to the facts of the case at bar. The cyclist, who was stopped and waiting to make a left-hand turn, was legally positioned in a primary traffic lane when he was struck from behind. *Taylor v. Bohemia, Inc.*, 688 P.2d 1374 (Or. Ct. App. 1984).

Pennsylvania

A twelve-year-old cyclist attempted to cross a roadway. He saw a vehicle some 750 feet in the distance and determined that he could make his way across the road in front of the vehicle. The motorist hit and severely injured the cyclist. The trial court judge entered a verdict for the cyclist. The Court of Appeals affirmed, holding that the evidence on the issue of the motorist's negligence supported the verdict. The motorist saw the young cyclist in the distance but continued at speeds estimated between sixty and sixty-five miles per hour. *Pierontoni v. Barber*, 119 A.2d 503 (Pa. 1956).

A twelve-year-old cyclist was riding with his older brother, on the shoulder of a highway, against traffic, at night, and without lights. An approaching motorist was driving on the shoulder. The older brother was able to leave the shoulder, but the twelve-year-old boy was hit and injured. The jury returned a verdict in favor of the cyclist, and the judgment was entered accordingly. The Supreme Court affirmed, holding that there was sufficient evidence to support the jury's conclusion that the cyclist was not contributorily negligent and that the cyclist's failure to have a light on his bicycle was not the proximate cause of the accident. *Masters v. Alexander*, 225 A.2d 905 (Pa. 1967).

A seventeen-year-old cyclist was riding at 11:30 in the evening on a high school driveway, without lights. A motorist turned onto the driveway, traveled some sixty to eighty feet, and collided with the cyclist. The jury returned a verdict in favor of the motorist. The Court of Appeals reversed, holding that it was error to instruct the jury that the cyclist was required to have an illuminated headlight in view of the evidence that the cyclist was operating his bicycle on the berm of the road. *Adams v. Mackleer*, 361 A.2d 439 (Pa. 1976).

A cyclist was riding down a road in Harrisburg and was hit by an overtaking motorist. The cyclist died from the injuries he sustained in the accident. After a jury trial, judgment was entered in favor of the motorist. The jury found the motorist's negligence to be twenty-one percent and that of the deceased cyclist to be seventy-nine percent. The Superior Court affirmed, holding that the testimony of the motorist that he observed the cyclist proceeding along a berm of highway on his bicycle in a straight and steady fashion, but that the cyclist suddenly appeared on the road in front of his automobile permitted submission of contributory negligence to jury despite the motorist's testimony that he had not kept his vision fixed exclusively on the cyclist. *Yandrich v. Radic*, 435 A.2d 226 (Pa. Super. Ct. 1981).

A cyclist, riding in the city streets of Philadelphia, was injured when he collided with a motorist pulling out of a parking spot. The motorist vacated the spot after looking back and seeing only a double-parked motorist waiting for the parking spot. After a jury verdict, the trial court entered judgment for the cyclist. The jury had found the cyclist forty percent contributorily negligent. The cyclist appealed. The Appellate Court affirmed, holding that there was evidence at trial of the cyclist's negligence to support the jury finding. The cyclist swerved around the double-parked vehicle and could have braked or swerved around the motorist. *Fox v. Waxman*, 1989 WL 147574 (E.D. Pa.).

A fifteen-year-old cyclist was riding at night without lights. He was riding in the middle of the right lane because the shoulder was partially rocky. He noticed the lights of an automobile approaching from behind. He braked and attempted to move from the lane, but was hit and injured by the motorist. The jury returned a verdict finding the cyclist eight-two and a half percent causally negligent, and the judgment was entered accordingly. The Superior Court reversed, holding that the cyclist's .06 percent blood alcohol level allowed no presumptive inference and that the cyclist was unfit to ride his bicycle. The presumptive level at law was .10%. *Locke v. Claypool*, 627 A.2d 801 (Pa. Super. Ct. 1993).

Rhode Island

A seventeen-year-old cyclist was riding his bike in the dark, early morning hours. A motorist traveling in the same direction as the cyclist hit him from behind. The jury returned a verdict in favor of the motorist. The trial court judge granted the cyclist's motion for a new trial The Supreme Court vacated, holding the trial court had misconceived material evidence. The court had applied the so-called "appellate rule" and looked at the evidence in the light most favorable to the motorist, the party that prevailed before the jury, and sustained their verdict as there was some competent evidence to support their finding. The trial judge relied on faulty reasoning on skid mark evidence. *Connors v. Gasbarro*, 448 A.2d 756 (R.I. 1982).

South Carolina

A cyclist was riding through an intersection in Greenville at around 5:20 on a February evening. A motorist pulled out from a stop sign and hit the rear of the cyclist's bike. The motorist testified that he never saw the cyclist. The trial court rendered judgment for the cyclist. The Supreme Court held that the motorist requested an instruction that if the cyclist's failure to have his bike equipped with lights was a proximate cause of the accident, the cyclist could not recover his damages was properly refused as an incomplete and erroneous proposition of law since it was not shown that lights were necessary. The court took judicial notice of the fact that the sun set on the day of the accident at least one hour after the accident. *Bradley v. Keller*, 156 S.E.2d 638 (S.C. 1967).

A fifteen-year-old cyclist was riding across attempting to enter a private driveway. He was hit by a motorist traveling on the road. The trial court judge directed verdict in favor of the motorist. The Court of Appeals reversed, holding that questions of whether the driver was negligent or reckless and whether this misconduct was the proximate cause of the accident were for the jury. The motorist had taken his eyes off of the road to wave to some people, did not brake or sound the horn. *Davenport v. Walker*, 313 S.E.2d 354 (S.C. Ct. App. 1984).

A cyclist was riding between the solid white line of highway and the dirt shoulder when he was struck from behind by a motorist. After the jury verdict, the trial court judge entered judgment for the motorist. The Court of Appeals reversed. Not allowing jury instruction on the "following too closely" statute was reversible error where the evidence at trial and pleading language of the complaint warranted the instruction. The cyclist was riding on the far right and was struck without warning. *Mouzon v. Moore & Stewart, Inc.*, 317 S.E.2d 756 (S.C. Ct. App. 1984).

A cyclist was riding down a well-lit two-lane highway at 4:20 a.m. when he was hit from behind by a motorist. The trial court judge entered judgment for the cyclist. The Court of Appeals held that the evidence supported a verdict for the cyclist. The cyclist heard the approaching motorist, did not look back and rode in a straight line. The jury award of punitive damages was warranted. The record contained evidence that the motorist violated either or both of the statutes relating to maximum speed limits and to overtaking and passing vehicles proceeding in the same direction. *Copeland v. Nabors*, 329 S.E.2d 457 (S.C. Ct. App. 1985).

A cyclist was riding down a City of Beaufort sidewalk (in violation of the city's bicycle ordinance) when he collided with a motorist exiting a private driveway. The trial court judge entered judgment in favor of the motorist. The Court of Appeals affirmed, holding that the cyclist's objection to the jury instruction concerning the city ordinance that renders it unlawful for any person . . . to ride a bicycle at any time on any of the sidewalks of the city was properly overruled. The city ordinance was not in conflict with the Uniform Act Regulating Traffic on Highways. *Burke v. Davidson*, 380 W.E.2d 839 (S.C. Ct. App. 1989).

South Dakota

An eleven-year-old cyclist and seven other boys were riding bicycles on the shoulder at night against traffic. A motorist traveling in the opposite traffic hit the eleven-year-old cyclist. The jury returned a verdict in favor of the cyclist, and the trial court judge entered the judgment. The Supreme Court affirmed, holding that it was for the jury to decide whether the eleven-year-old cyclist riding at night without lights, on the wrong side of the highway, was contributorily negligent. It was also for the jury to determine the factual issue of whether or not the motorist was negligent in failing to keep a proper lookout and to have her automobile under sufficient control in order to avoid the accident. *Finch v. Christensen*, 172 N.W.2d 571 (S.D. 1969).

Tennessee

An eight-year-old, riding on the handlebars of his brother's bicycle, was injured when they rode out from a residential driveway and collided with a motorist. The trial court judge rendered judgment for the motorist. The Court of Appeals held that the minor cyclist had the duty to yield right-of-way to the motorist. The motorist had managed to bring her vehicle to a complete stop upon seeing the children entering the roadway and was not negligent. *James v. Ross*, 369 S.W. 2d 1 (Tenn. 1962).

A twelve-year-old cyclist was riding on the wrong side of a two-lane road and was hit by an oncoming motorist. The trial court entered judgment for the cyclist. The Court of Appeals affirmed, holding that the evidence supported the conclusion that the northbound motorist, who entered the curve at a speed between forty-five and fifty miles per hour and who left sixty-two yards of skid marks as his auto left the road in an attempt to avoid oncoming southbound vehicles and the southbound cyclist riding in the wrong lane of traffic, was guilty of negligence that proximately caused or contributed to the accident. *Morris v. Summers*, 474 S.W.2d 662 (Tenn. Ct. App. 1971).

A cyclist, riding on the shoulder against traffic, was hit by a motorist pulling out from a stop sign. The trial court entered judgment for the cyclist after the jury verdict. The Court of Appeals reversed, finding that the evidence at trial preponderated against finding that the cyclist was only remotely contributorily negligent and in favor of finding that the cyclist's negligence was the proximate cause of his own injuries, thereby barring his recovery. *Grazer v. Windham*, 640 S.W.2d 27 (Tenn. Ct. App. 1982).

A fifteen-year-old boy was riding with friends in a church parking lot. He rode out into a lane of traffic and was hit and killed by a motorist. The trail court eventually entered judgment against the motorist. The Court of Appeals reversed, holding that the motorist was not liable for the inattentive cyclist's death where the evidence was overwhelming that the driver did not see the cyclist, who had placed himself in a position of peril, in time to avoid the accident. The cyclist was one and a half car lengths in front of the motorist when she first saw him. *Braden v. Hall*, 730 S.W.2d 329 (Tenn. Ct. App. 1987).

Texas

A thirteen-year-old cyclist was riding with a friend one-half hour after sunset. They came to a crossroad. After stopping and looking both ways, the thirteen-year-old started across the road and was hit by a truck. The trial court judge entered the verdict in favor of the motorist after the jury found both parties negligent. The Court of Appeals reversed, holding that the jury's finding of negligence on the cyclist's part was against the great weight and preponderance of the evidence. The trucker was traveling at a high, excessive and negligent rate of speed. There was no direct evidence that the cyclist failed to keep a proper lookout, nor that the lookout he kept was the proximate cause of the accident. *Kelly v. Hamm-Tex Distributing Company*, 337 S.W.2d 608 (Tex. App. 1960).

A thirteen-year-old cyclist riding on a two-way street, against traffic, came to a one-way cross street. He saw an approaching motorist, but believed he could make it across in time. He started across and was hit by the motorist. The trial court entered judgment for an instructed verdict. The Court of Appeals reversed, holding that the evidence that the cyclist entered the intersection after stopping and looking both ways was free of negligence. The motorist was speeding and failed to maintain a proper lookout. *Pickford v. Broady*, 411 W.W.2d 747 (Tex. App. 1966).

A cyclist came to an intersection. After a complete stop, the cyclist entered the intersection and was hit by a motorist. The trial court judge entered the verdict in favor of the cyclist. The Court of Appeals affirmed, holding that the undisputed evidence at trial established that the motorist was in the intersection when he collided with the cyclist, and the only dispute was how far into the intersection he had gone. His failure to yield to the right-of-way was not in dispute, and the trial court did not err in submitting special issue that assumed failure to yield right-of-way by the motorist. *Hughett v. Dwyre*, 624 S.W.2d 401 (Tex. App. 1981).

A cyclist was descending down Red Bud trails in West Lake Hills, west of Austin. A motorist, heading in the opposite direction, was waiting to take a left-hand turn and did not see the cyclist. The motorist turned in front of the cyclist. The trial court judge entered a take-nothing judgment against the cyclist. The Court of Appeals affirmed, holding that the instruction on negligence per se based on the violation of the left-turn statute was not warranted and had been properly rejected. The mere fact that the motorist turned in front of the cyclist without proof offered that doing so was a failure to act as a reasonably prudent person under the circumstances, did not warrant the negligence per se instruction. The testimony of the accident reconstructionist was properly admitted. There was no abuse in discretion in admitting expert testimony finding that the motorist had not been negligent was supported by evidence. The jury's finding after review of the photographs, competing testimony was not against the great weight and preponderance of the evidence. The cyclist was wearing dark cycling cloth that blended into the trees behind him as he descended the hill. *Waring v. Wommack*, 945 S.W.2d 889 (Tex. App. 1997).

Utah

A fifteen-year-old developmentally disabled cyclist failed to stop at a stop sign and was hit by a motorist. The trial court judge entered judgment for the cyclist. The Supreme Court affirmed, holding that the jury was properly instructed on the last clear chance doctrine. The evidence offered by the cyclist's expert at trial showed that the motorist was traveling at thirty-five miles per hour and had between five and seven seconds from the time he saw the cyclist to avoid the accident. *Reese v. Proctor*, 487 P.2d 1267 (Utah 1971).

A cyclist was riding down hill at night without lights. A motorist, traveling in the opposite direction, turned left in front of her, and the cyclist collided with her vehicle. The jury returned a verdict in favor of the motorist, and the judgment was entered accordingly. The Court of Appeals affirmed, holding, inter alia, that the jury finding that the motorist was negligent, but that such negligence was not the proximate cause of the accident, was supported by the evidence. *Rasmussen v. Sharapata*, 895 P.2d 391 (Utah Ct. App. 1995).

Vermont

An eight-year-old cyclist on his way home from school for lunch rode from a path onto the roadway. He saw a motorist approaching and noticed that the driver was not looking in his direction. He tried to stop but then decided to pedal out of the way. He was hit by the motorist. After a jury verdict for the motorist, the trial court judge entered judgment in her favor. The Supreme Court affirmed, holding that although the motorist is required to maintain a constant watch for persons and property on a highway, the motorist is not necessarily negligent merely because she fails to look ahead continually and uninterruptedly. The motorist had taken her eyes off of the rode between two and three seconds to look to her left to look for her daughter when the collision occurred. *Beaucage v. Russell*, 238 A.2d 631 (Vt. 1968).

Virginia

A twelve-year-old cyclist turned between two parked vehicles to avoid traffic. A motorist had placed large sharp mower blades in his truck, which extended between eighteen and twenty inches form the end of the truck. The boy rode into the blades and severely injured his arm. The jury found for the cyclist, and the trial court judge entered the verdict accordingly. The Supreme Court affirmed, holding that the motorist is liable for all consequences that naturally flow from his negligence. Foreseeability of a particular danger is not required. The cyclist was not contributorily negligent. *Judy v. Doyle*, 108 S.E. 6 (Va. 1921).

A seven-year-old cyclist was riding on the rear fender of another minor cyclist's bicycle when they were hit by a turning motorist who had not seen them. The jury returned a verdict in favor of the motorist, and the judgment was entered accordingly. The Supreme Court reversed, holding it was reversibly erroneous for the trial court to instruct that the riding of more than one person on one bike is a violation of the statute. *Phillips v. Schools*, 175 S.E.2d 279 (Va. 1970).

A thirteen-year-old cyclist was riding with a companion side-by-side on the shoulder of a road. They were startled by an overtaking motorist and attempted to cross to the other side of the road. The thirteen-year-old was hit by the motorist. After the cyclist's case in chief, the trial court judge directed a verdict for the motorist. The Supreme Court reversed, holding that the evidence as to whether the motorist kept a proper lookout, increased her vigilance, and had her car properly under control when she attempted to pass the cyclists was for the jury. She testified that she could not tell if the cyclists were heading in her direction or away from her. She attempted to pass them at fifty miles per hour. *Endicott v. Rich*, 348 S.E.2d 275 (Va. 1986).

A fourteen-year-old cyclist was riding his bike down a hill on the way to school at approximately 7:00 on an October morning. A truck turned left in front of him at an intersection, and the cyclist collided with the truck. The trial court judge granted summary judgment for the motorist. The Supreme Court reversed, holding that the cyclist's violation of the statute requiring bicycles to be equipped with an operable headlamp between sunset and sunrise did not preclude, as a matter of law, the cyclist's action against the motorist. Reasonable minds could aver whether the cyclist's violation of the statute was a proximate cause of the damages sustained by the cyclist. *Karim V. Grover*, 369 S.E.2d 185 (Va. 1988).

A cyclist was riding in one of two southbound lanes. As an overtaking motorist was passing, the cyclist fell and died from his injuries. The jury returned a verdict in favor of the cyclist. The trial court judge set it aside and entered judgment for the motorist. The Supreme Court reversed in regards to the motorist's negligence, holding that the evidence supported the jury's finding that the cyclist's injuries and death were the proximate result of the motorist's driving and passing too closely to the cyclist. *Mann v. Hinton*, 457 S.E.2d 22 (Va. 1995).

Washington

A fourteen-year-old was riding on the shoulder of a road when an overtaking motorist sounded his horn as he passed the cyclist. Just as the motorist was passing, the cyclist veered left directly in front of the motorist. He was hit and killed. The jury found both the motorist and the cyclist negligent, and the trial court judge entered judgment for the motorist under Washington's contributory negligence statutes. (The applicable law at the time of the accident.) The Supreme Court affirmed, holding that the doctrine of last clear chance was not applicable as to the motorist whose overtaking automobile struck and killed the overtaken cyclist, where the cyclist rode to the shoulder of the road as the motorist sounded his horn while passing at forty miles per hour in a twenty-five miles per hour speed zone and the motorist was still exceeding the speed limit when the cyclist abruptly rode directly in front of the automobile as it was passing the cyclist. The motorist did not have time to see the peril the cyclist was in and to appreciate the danger and effectually act. *Radecki v. Adams*, 387 P.2d 974 (Wash. 1964).

A cyclist and a truck came to the same intersection at about the same time. The cyclist was making a right from a stop sign, and the motorist was making a left. The motorist cut the corner and hit the cyclist. The trial court granted the motion for summary judgment on liability against the motorist. The Court of Appeals affirmed, holding that where the driver of the truck had not only crossed the center lines of the intersecting streets while turning left, which in itself was violative of a statute, but had taken up almost all of the side of the road on which the cyclist was permitted to travel, the truck driver was negligent; that the truck driver's acts were the proximate cause of the collision; and, that the cyclist who allegedly failed to stop at the stop sign was not contributorily negligent. *Foster v. Bylund*, 503 P.2d 1087 (Wash. Ct. App. 1972).

A cyclist was descending down a hill on Mission Avenue in Spokane. The cyclist saw a truck at a cross road stop sign and took his eyes off of the truck for between five and seven seconds. When he looked up, the truck was in his lane, and he collided with it. The trial court judge found the cyclist negligent in failing to yield the right-of-way and negligent in failing to maintain a proper lookout. The Court of Appeals reversed, holding that the cyclist was not guilty of contributory negligence on the theory that he failed to maintain a proper lookout, and he was entitled to recover for injuries received in the collision. The cyclist had every reason to believe that the motorist would continue yielding the right-of-way, as he was legally obligated to do. *Merrick v. Stansbury*, 533 P.2d 136 (Wash. Ct. App. 1975).

A minor cyclist stepped from the curb into the crosswalk after a truck stopped to allow her to cross. When she was approximately midway into the crosswalk, in front of the truck, a motorist waiting behind the truck pulled around, accelerated and hit the young cyclist. The jury returned a verdict in favor of the cyclist, but also finding her fifty percent contributorily liable. Because of instructional error, the trial court ordered a re-trail. The motorist appealed. The Court of Appeals affirmed, holding that the motorist was governed by a statute, which provides that the driver of a vehicle approaching from the rear shall not overtake and pass another vehicle stopped at a crosswalk in order to permit a pedestrian to cross the roadway. The mounting of the bicycle by the cyclist in the crosswalk had no bearing on the situation. The crosswalk is not a roadway. *Crawford v. Miller*, 566 P.2d 1264 (Wash. Ct. App. 1977).

Two cyclists, ages thirteen and fourteen, were riding in a westerly direction on a two-lane undivided highway. An approaching motorist honked his horn to warn of his approach. The fourteen-year-old crossed to the eastbound lane and continued westbound. Just as the motorist was passing between the cyclists, the thirteen-year-old boy, who was startled by the motorist, feel in the path of the motorist and was injured. The trial court judge directed, as a matter of law, that the injured cyclist was contributorily negligent and that such negligence was the proximate cause of his injuries. The Supreme Court held that the evidence was for the jury as to the cyclist's contributory negligence. The trial court's directing issues was not harmless error. The jury could have reasonably concluded that the cyclist was startled by the approaching car and that this was the proximate cause of the accident. *Bordynoski v. Bergner*, 644 P.2d 1173 (Wash. 1982).

A nine-year-old cyclist rode into the street and was hit by a motorist. The trial court judge entered summary judgment in favor of the motorist The Court of Appeals affirmed, holding that the plaintiff did not state sufficient facts to take her case to the jury since, even accepting the expert's evidence that the motorist was going forty-two miles per hour and that she was fifty-nine feet back from the child cyclist when she first saw him, the collision, nevertheless, would have occurred even at the legal thirty-five miles per hour speed limit. Therefore, the evidence would not support a finding that there was any negligent act of defendant that was a proximate cause of the accident. *Theonnes v. Hazen*, 681 P.2d 1284 (Wash. 1984).

An eleven-year-old paperboy was riding his bike in a cross walk and was hit by a motorist. The trial court judge entered judgment on the jury's special verdict finding no negligence on the part of the motorist. The Court of Appeals affirmed, stating that there was no error in its refusal to instruct on the cyclist's proposed instruction regarding the "duty to see." The general negligence instruction adequately allowed the cyclist's theory of the case in that it defined negligence in terms of both the doing of an act and the failure to do an act. The trial court's general negligence instruction did not unfairly emphasize the cyclist's "duty to see" at the expense of the motorist's "duty to see." *Young v. Carter*, 684 P.2d 784 (Wash. Ct. App. 1984).

A fourteen-year-old cyclist was riding down a steep hill on Seattle Street, at night, without lights, and collided with a left-turning motorist. The jury found for the cyclist, but found him ninety-five percent causally negligent in the accident. The Court of Appeals affirmed. The Supreme Court reversed, holding that the violation of the statute by minors, between the ages of six to sixteen, does not constitute proof of negligence per se, but may, in proper cases, be introduced as evidence of the minor's negligence. *Bauman v. Crawford*, 704 P.2d 1181 (Wash. 1985).

A cyclist, riding on the shoulder, was overtaken by a motorist who then turned right. The cyclist collided with the vehicle and was injured. The jury returned a verdict finding that the motorist was not negligent. The Court of Appeals affirmed, holding that even if it was error to give "following vehicle" instruction, the instruction was harmless error because it pertained only to the contributory negligence of the cyclist, and, therefore, the instruction was not considered by the jury when the jury found the motorist free of negligence. The cyclist had argued that the instruction did not apply to a cyclist riding on the "shoulder" in the statute that applies rules of the road to bicycles. Because the court determined the instruction was not prejudicial, it did not determine whether it was an error to give it. *Sanders v. Stuart*, 1998 WL 703327 (Wash. App. Div. 2).

A cyclist was riding south on the Interurban bicycle trail in Kent which intersects 277th Street at a marked crosswalk. The cyclist stopped at the stop sign. A westbound motorist stopped to allow him to cross, but an eastbound motorist struck him. The trial court granted summary judgment to the cyclist. The Supreme Court affirmed, holding that once the cyclist entered the crosswalk, he had the right-of-way over the approaching motorist. A crosswalk user on a bicycle is entitled to the same protections and has the same duties as a crosswalk user on a skateboard, on foot, or in a wheelchair, and once a bicyclist safely enters the crosswalk, he or she has the right-of-way over an approaching motorist. The court applied the motorist's interpretation of the applicable statute that cross walk protections do not extend to persons riding bicycles in the following example. Several groups of children return home from school by the interurban Trail, some on foot, others on skateboards, roller blade, or bicycles, and wait at the crosswalk for a clear opportunity to cross 277th Street, and, like the cyclist at bar, they proceed only after the traffic has stopped for them, and after they have properly checked for oncoming traffic. If such a group were hit in the crosswalk by a motorist, the vehicle's driver would be liability to all the children except those on bicycles. The court called such a result nonsensical. *Pudmaroff v. Allen*, 977 P.2d 574 (Wash. 1999).

West Virginia

A ten-year-old cyclist was riding on the right. He was hit by an overtaking motorist. The trial court entered judgment in favor of the cyclist. The Court of Appeals affirmed, holding that the testimony of the sheriff, based on physical facts found at the accident scene, that the automobile struck the cyclist directly from behind in the center of the right lane of the highway, was not conclusive as to the cause of the accident, and the testimony was relevant and competent. The court rejected the motorist's characterization of the sheriff's opinion testimony as being conclusive as to the cause of the accident. *Jordan v. Bero*, 210 S.E.2d 610 (W. Va. 1974).

A fourteen-year-old was riding in the right lane of traffic when he was approached form behind by a motorcycle. The cyclist moved left, and the motorcyclist passed the cyclist on the right. The cyclist collided with a motorist traveling in the opposite direction. The trial court judge granted summary judgment to the motorcyclist. The Supreme Court reversed, holding that the material issue of fact as to whether the motorcycle passed the bicyclist on the right side of the road in violation of a statute, forcing the cyclist into the accident with the automobile passing in the opposite direction. There existed some uncertainty regarding the exact manner and sequence of the interaction between the cyclist and the motorcyclist. *Sartin v. Evans*, 414 S.E.2d 874 (W.Va. 1991).

Wisconsin

An eight-year-old was learning to ride a bike. He was hit by a motorist traveling in the opposite direction. The trial court entered the verdict in favor of the cyclist. The Supreme Court held that while the motorist was not an insured of the young cyclist's safety, he had a duty to operate his vehicle as closely as possible to the right-hand edge or curve of the roadway and that the responsibility for the accident was upon the motorist for entering wrong lane and colliding with the cyclist. The motorist testified that he saw a small child "zigzagging" on a bike and did nothing to decrease his speed when he could have stopped within four to six feet. *Leiner v. Kohl*, 52 N.W.2d 154 (Wis. 1952).

A ten-year-old cyclist was riding with two other young cyclists. They heard a truck approaching from behind, and two of them went onto the sidewalk. The ten-year-old cyclist was hit and killed by the overtaking truck. The roadway was narrowed by snow banks and parked cars. The judge entered the judgment for the cyclist after the jury found the motorist to eighty percent negligent, the parked car ten percent, and the cyclist ten percent. The Supreme Court affirmed, modifying the assigned negligence to be one hundred percent upon the motorist. There was no evidence that the owner of the parked car violated any duty. The cyclist was riding far too the right, had no duty to overtaking the motorist, and the motorist was the sole cause of the accident, by failing to see or failing to give more room in passing. *Fox v. Pettis*, 114 N.W.2d 836 (Wis. 1962).

A cyclist was riding to the west (two to three feet) of a line of parked cars in Oconomowoc. A parked driver opened her car door, causing the cyclist to veer left and be hit by an overtaking motorist. The jury returned a verdict, finding the "door opener" one hundred percent at-fault and finding no fault upon the overtaking motorist. The Supreme Court held that the evidence that the motorist was too "stunned" to stop immediately and dragged the cyclist for a distance of fifty-eight feet made a question for the jury of whether she conformed to the standard of reasonable or prudent person placed in like circumstances. A jury question was presented as to whether the motorist's failure to stop in a shorter distance, considering that her vehicle was dragging the cyclist and that this contributed to the severity of the injuries, was in itself a separate item of negligence. *Zillmer v. Miglautsch*, 151 N.W.2d 741 (Wis. 1967).

An eight-year-old cyclist rode behind a dump truck and was killed when the truck backed up. The trial court judge entered judgment in favor of the motorist after the jury found the cyclist to be ninety percent negligent and the motorist to be ten percent negligent. The Supreme Court affirmed, holding that a rearview mirror would not have disclosed the presence of someone as close as the deceased cyclist in the instant case and that any alleged violation of the rearview mirror statute could not be considered a causal factor in the accident. *Grube v. Moths*, 202 N.W.2d 261 (Wis. 1972).

A cyclist was riding west, and as he entered an intersection, a motorist heading east turned left in the cyclist. The trial court entered judgment in favor of the cyclist. The Court of Appeals affirmed, holding that the statute, providing the pedestrian or bicyclist crossing the roadway at a point other than at the crosswalk must yield to vehicles using the roadway was not applicable to automobile-bicycle collisions. The vehicle attempting to turn left was required to yield the right-of-way to the oncoming cyclist. *Chernestski v. American Family Mutual Insurance Company*, 515 N.W.2d 283 (Wis. Ct. App. 1994).

A cyclist was hit by a City of Milwaukee fire truck, which made a right turn into a private driveway. The jury found the motorist ninety percent causally negligent. The Court of Appeals affirmed, holding that the City had no basis for its argument that the cyclist was riding on the sidewalk. The cyclist, who was on a racing bike, in his riding clothing, on a seven-mile commute, testified that he never rode on the sidewalk, and witnesses agreed. The court went on to say that it made little difference in the merits of the suit whether the cyclist was riding on the road or the sidewalk. The fire truck would have to yield the right-of-way in either case. *Alioto v. Oglesby*, 532 N.W.2d 144 (Wis. Ct. App. 1995).

Wyoming

A nine-year-old cyclist was riding on a highway. An overtaking motorist collided with the young cyclist. The trial court judge entered the verdict for the cyclist. The Supreme Court affirmed, holding that the evidence supported the finding that the motorist, who was traveling in the same direction on the highway, was guilty of negligence in passing the cyclist. The trial judge could well have given to the testimony of the motorist the weight and consideration required and still have arrived at a finding of negligence on the part of the motorist. The motorist did not slow as he passed. The cyclist was not contributorily negligent. *Blakeman v. Gopp*, 364 P.2d 986 (Wyo. 1961).

Engineering Appendix A

Railroad Information

Part I
Guidelines for the Construction or Reconstruction of Highway-Railway Crossings
American Railway Engineering Association
(1987)

1.0 Crossing Surface Materials

Any crossing surface material may be used on any crossing at the discretion of the operating railroad or as may be recommended by a diagnostic evaluation of the crossing.

Specifications and plans concerning the crossing surface material and use should abide with the manufacturer's recommendations, and/or the operating railroad's specifications and plans and, where applicable, to the standards of the public agency having jurisdictional authority at the specific location.

1.1 Width of Crossing

The crossing shall be of such width as prescribed by law, but in no case shall the width be less than that of the adjacent traveled way plus 2 ft.

1.2 Profile and Alignment of Crossings and Approaches

Where crossings involve two or more tracks, the top of rails for all tracks shall be brought to the same plane where practicable. The surface of the highway shall be in the same plane as the top of rails for a distance of 2 ft. outside of rails for either multiple or single-track crossings. The top of rail plane shall be connected with the grade line of the highway each way by vertical curves of such length as is required to provide riding conditions and sight distances normally applied to the highway under consideration. It is desirable that the surface of the highway be not more than 3 in. higher nor 6 in. lower than the top of nearest rail at a point 30 ft. from the rail, measured at right angle thereto, unless track superelevation dictates otherwise.

If practicable, the highway alignment should be such as to intersect the railroad track at or nearly at right angles.

1.3 Width and Marking of Approaches

Width of roadway at a highway-railway grade crossing should correspond to that of the adjoining highway and have the same number and width of traffic lanes as adjoining highway without extra lanes and with center turn lanes at the crossing being delineated.

At all paved approaches to the highway-railway grade crossing, the highway traffic lanes in the vicinity of the crossing should be distinctly marked in accordance with the recommendations of the Manual on Uniform Traffic Control Devices for Streets and Highways. Such markings are the responsibility of the public authorities.

1.4 Drainage

In situations where the grade of the highway approach descends toward the crossing, provisions shall be made to intercept surface and subsurface drainage and discharge it laterally so that it will not be discharged on the track area.

Surface ditches shall be installed. If required, subdrainage with suitable inlets and the necessary provisions for clean-out shall be made to drain the subgrade thoroughly and prevent the formation of water pockets. This drainage shall be connected to a storm water sewer system, if available; if not, suitable piping, geotextile fabrics and/or french drains shall be installed to carry the water a sufficient distance from the roadbed. Where gravity drainage is not available, a nearby sump may provide an economical outlet, or the crossing may be sealed and the roadbed stabilized by using asphalt ballast or its equivalent.

1.5 Ballast

The ballast and sub-ballast shall be dug out a minimum of 10 in. below the bottoms of the ties, 1 ft. minimum beyond the ends of the ties and beyond the end of the crossing a minimum of 20 ft., and reballasted with ballast to conform with AREA specifications.

1.6 Ties

Treated No. 5 hardwood or concrete ties shall be used through the limits of the crossing and beyond the crossing a minimum of 20 ft.

1.7 Rail

The rails throughout the crossing shall be so laid to eliminate joints within the crossing. Preferably, the nearest joint should be not less than 20 ft. from the end of the crossing. Where necessary, long rails shall be used or the rail ends shall be welded to form continuous rail through the crossing. Rails shall be spiked to line, and the track shall be thoroughly and solidly tamped to uniform surface. Rails should be protected with an approved rust inhibitor.

1.8 Flangeway Widths

Flangeways not less than 2½ or more than 3 in. width should be provided. Flangeways shall be at least 2 in. in depth unless approved by the operating railroad.

1.9 New or Reconstructed Track Through Crossing

1.9.1 Profile

An agreed upon profile, railroad and highway, should be established between the operating railroad and the road authority.

1.9.2 Subgrade

Subgrade should be cleaned of all old contaminated ballast and bladed to a level surface at a minimum of 10 in. below bottom of tie and at least 20 feet beyond each end of the crossing.

1.9.3 Drainage Areas

All drainage areas should be cleaned and sloped away from the crossing both directions along the track and the roadway. See Paragraph 1.4 above.

1987

1.9.4 Geotextile Fabrics

When practical, a geotextile fabric should be used between the subgrade and the ballast section and at least 20 ft. beyond each end of the crossing and if a rail joint falls within these limits at least 5 ft. beyond the rail joint. If practical, the geotextile fabric should extend under the roadway surface, traveled way, 15 ft. each way from center line of track.

1.9.5 Ballast

A minimum of at least 10 in. of clean ballast should be placed between bottom of tie and the sub-ballast or subgrade. See Paragraph 1.5 above.

1.9.6 Ties

No. 5 hardwood or concrete ties should be used. Length and spacing of ties should conform to the type of grade crossing surface materials being used. See Paragraph 1.6 above.

1.9.7 Tie Plate, Spikes, Anchors

All ties through the crossing area and at least 20 feet beyond each end of the crossing should be fully tie plated with four spikes per tie plate and fully box anchored. Optional placement of the tie pads is acceptable.

1.9.8 Rail

The rail section should, at a minimum, be 115 lb. welded rail throughout the crossing to at least 20 feet beyond each end of the crossing.

1.9.9 Lining and Surfacing Track

Rails should be spiked to line and the track machined or mechanically tamped and surfaced to grade and alignment of the existing track and roadway as described in Paragraph 1.2 above. Let as many train movements, as time will permit, across the crossing before final surface and alignment. This will help achieve the optimum ballast compaction through the crossing area.

1.10 Highway Approaches

The width, transverse contour, type of surface or pavement, and other characteristics of each such approach to a high-railway grade crossing should be suitable for the highway and the railroad and shall, in each case, conform to the requirement of good practice.

1.11 Highway Traffic Control in Railroad Work Zones

Traffic control is recommended when a maintenance or construction activity interferes with vehicle traffic flow. Federal guidelines for the devices and methods of traffic control are outlined in the *Traffic Control Devices Handbook*, published by the Federal Highway Administration. However, since State and/or local practices and guidelines will vary, coordination between the railroad(s) and the appropriate roadway authority is required.

EDITOR'S NOTE

The Manual on Uniform Traffic Control Devices for Streets and Highways referred to in this chapter was prepared by the National Advisory Committee on Uniform Traffic Control Devices for Streets and Highways (now organized as the National Committee on Uniform Traffic Control Devices for Streets and Highways). The Manual may be purchased from the Lawyers & Judges Publishing Co.

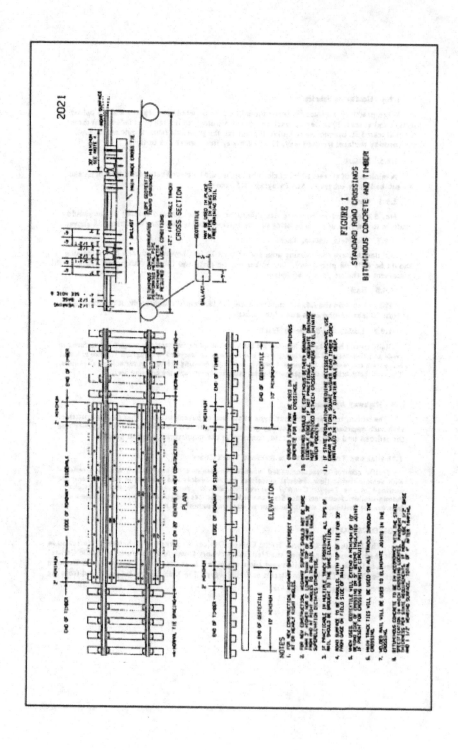

Engineering Appendix A - Railroad Information

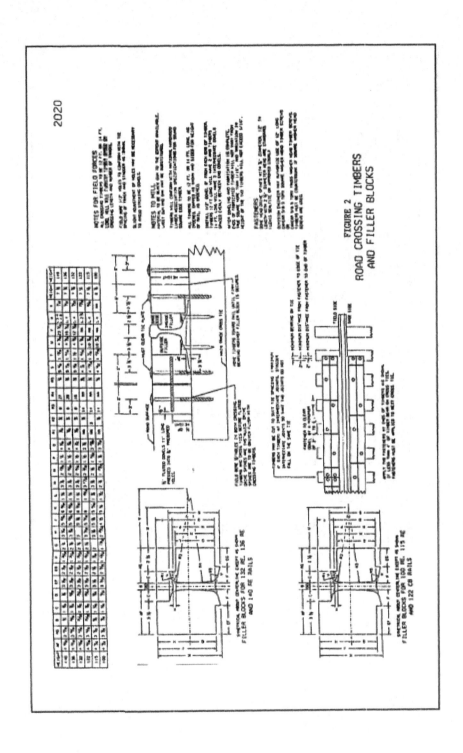

FIGURE 2
ROAD CROSSING TIMBERS AND FILLER BLOCKS

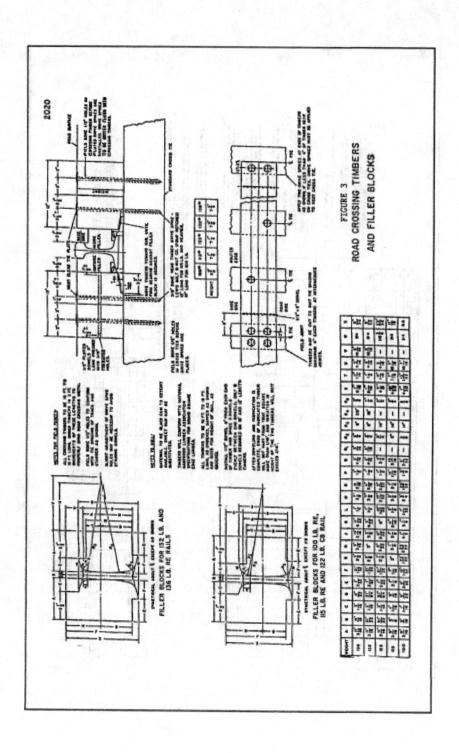

FIGURE 3
ROAD CROSSING TIMBERS AND FILLER BLOCKS

Engineering Appendix A - Railroad Information

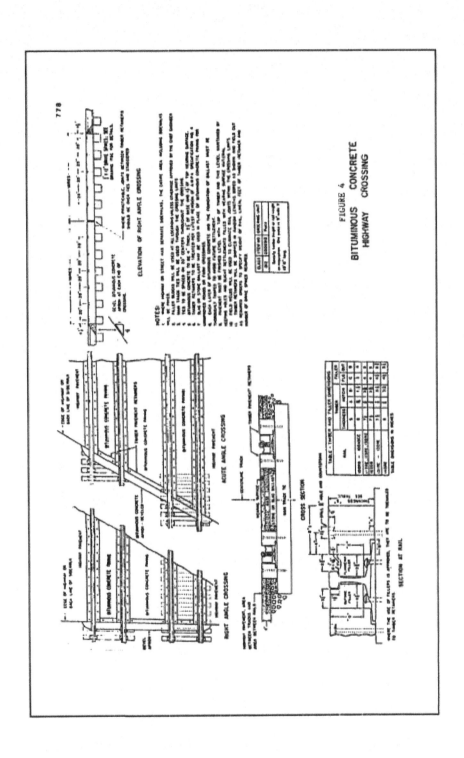

FIGURE 4
BITUMINOUS CONCRETE HIGHWAY CROSSING

Engineering Appendix B

Appropriate Case Citations

Summary of Case Citations

The selected case citations fall within several general subject areas. In the brief discussion of these general areas which follows, the citations are referred to only by plaintiff last name. These references should, however, be sufficient to aid the reader's search for further detail.

Bicycle collisions represent a large number of the legal actions in which bicycle accident reconstruction is employed. Those citations which involve a bicycle collision with an automobile or truck include Bennett, Johnston, Laird, McWhorter, Moore, Ralston, and Smith. A bicycle/pedestrian collision underlies the Dennis case.

Site related cases address the liability of the site owner. A railroad crossing case is detailed in Schaffner. For cases addressing the liability of a park, city, or private property owner, one should read Barnes, Bledsoe, Dennis, and Walker.

The validity of executed liability releases or waivers has been addressed in several cases. These cases include Bennett, Buchan, Castro, Hiett, and Madison.

Various issues arising from accidents involving a minor child cyclist are discussed in Barnes, McWhorter, Moore, Ralston, and Smith. Ralston, together with Johnston, also deal with the issue of investigating officer or witness testimony. Further, Smith confronts allegations of wanton misconduct by a law enforcement officer in pursuit of a third party.

Defective, unsuitable, or improperly assembled bicycle components and accessories are the basis for many bicycle accident lawsuits. Two cases involving quick release front wheel mechanisms are Schaffner and Swanlund. Caporale deals with improper assembly. The bicycle light is at issue in McWhorter. Nonuse of or inadequate helmet is addressed in Moore and Buchan.

The issue of the cyclist's reasonable standard of care is discussed in Buchan, Johnston, and Moore. Bicycle manufacturer or distributor liability is found in Caporale and Swanlund. The citations provided in this book are only a sample of the available cases which involve bicycle and related accidents and encompass a large number of the more common types of bicycle accident lawsuits.

Cite as 760 P.2d 566 (Ariz. App. 1988)
157 Ariz. 566
Nancy K. BARNES, as lawful custodian, guardian and
conservator of Marc R. Barnes, a minor,
Plaintiff/Appellant,
v.
CITY OF TUCSON, DefendarWAppellee.
No. 2 CACV 87-0297.
Court of Appeals of Arizona, Division 2, Department B.
February 29,198s.
Reconsideration Denied March 29,1988.
Review Denied September 14,198s.

Mother brought action against city for injuries juvenile bicyclist suffered when he collided with post erected by city on edge of city sidewalk. The Superior Court, Pima County, Cause No. 230011, John F. Kelly, J. pro tem, entered judgment for mother, and she appealed. The Court of Appeals, Fernandez, J., held that: (1) duty owed by city to public was to keep its streets and sidewalks reasonably safe; (2) instruction that violation of ordinance prohibiting operation of bicycle upon public sidewalk was negligence per se was properly given, despite claim that ordinance was enacted to protect pedestrians and city was not within class of pedestrians designed to be protected by it; and (3) new trial was required on damages, where admitted medical bills totaled $6,707.83, there was no contrary evidence on medical expenses, and damage award was $3,000.

Affirmed; remanded on issue of inadequacy of damages.

1. Duty owed by city to public is to keep its streets and sidewalks reasonably safe; the city's duty is not governed by rules pertaining to negligence actions against possessors of land.

2. In action for injuries suffered by juvenile bicyclist when he collided with post erected by city on edge of city sidewalk, instruction that violation of ordinance prohibiting operation of bicycle upon public sidewalk was negligence per se was properly given, although it was claimed ordinance was enacted to protect pedestrians and city was not within class of persons designed to be protected by it; the ordinance is intended not only to protect pedestrians from bicycle riders, but also to protect bicycle riders from hazards of riding on narrow sidewalks.

3. For purposes of finding negligence per se on part of juvenile bicyclist

injured when he collided with post erected by city on edge of city sidewalk, based on juvenile's violation of ordinance prohibiting operation of bicycle upon public sidewalk, ordinance applies to all persons riding bicycles, including minors, despite claim that minor who was not engaged in adult activity could not be guilty of negligence per se and that it was error to instruct on negligence per se since riding bicycle was not typical adult activity; State traffic laws on riding of bicycles apply to all persons, regardless of age. A.R.S. V 28-218.

4 . In action for injuries suffered by juvenile bicyclist who collided with post erected by city on edge of city sidewalk, fact that juvenile was minor was properly taken into account in instructing jury, where instructions on standard of care applicable to child and instruction on city's duty to anticipate behavior of children were given, as well as instruction that violation of ordinance prohibiting operation of bicycle upon public sidewalk was negligence per se.

5 . New trial would be ordered on issue of damages awarded in action for injuries suffered by juvenile bicyclist who collided with post erected by city on edge of city sidewalk; city stipulated to admission of medical bills, which totaled $6,707.83, there was no contrary evidence on medical expenses, but juvenile was awarded damages of only $3,000, and court stated at hearing on new trial motion that it thought jury was confused about damages and might have misconstrued the instructions as there was no other explanation for the $3,000 award that the court could think of.

Zlaket & Zlaket, PC. by Eugene Zlaket, Tucson, for plaintiff/appellant.

Hazlett & Wilkes by Carl E. Hazlett and Jay Felix, Tucson, for defendant/appellee.

Cite as 239 Cal. Rptr. 55 (Cal. App. 2 Dist. 1987)
193 Cal. App. 3d 1485
Albert I. BENNETT, Plaintiff and Appellant,
UNITED STATES CYCLING FEDERATION
et al., Defendants and Respondents, Civ. B022865
Court of Appeal, Second District
Division 5.
August 4, 1987

Amateur bicyclist brought personal injury action following collision with automobile during race. Defendants moved for summary judgment. The Los Angeles Superior Court, H. Walter Croskey, J., granted motion and dismissed complaint. On appeal, the Court of Appeal, Epstein, J., assigned, held that:

(1) release was not invalid because of print size, and

(2) genuine issue of material fact as to whether presence of automobile on barricaded course was obvious or reasonably foreseeable hazard contemplated by bicyclist, who had signed release, precluded summary judgment.

Reversed.

3. Print size is important factor, but not necessarily the only one to be considered in assessing adequacy of document as release.

4. Release signed by amateur bicyclist prior to race was not invalid because of its five and one-half type print size; release was sufficiently conspicuous and legible, since release language was practically the only language on document and did not have to compete with other less important information for the subscriber's attention.

5. To be effective, written release purporting to exculpate tort-feasor from future negligence or misconduct must be clear, unambiguous, and explicit in expressing intent of subscribing parties.

6. Genuine issue of material fact, as to whether presence of automobile on barricaded bicycle race course was obvious or reasonably foreseeable hazard of type contemplated by bicyclist signing release from liability, precluded summary judgment in bicyclist's personal injury action following collision.

Cite as 823 P2d 1264 (1991)
170 Ariz. 256
Joseph K. BLEDSOE, a single man, Petitioner,
v.
The Honorable Stanley Z. GOODFARB, Judge of the Superior Court of the State of Arizona, in and for the County of Maricopa, Respondent,
and
SALT RIVER VALLEY WATER USERS ASSOCIATION, an Arizona corporation, Real Party in Interest.
SALT RIVER VALLEY WATER USERS ASSOCIATION, Petitioner,
v.
The Honorable Stanley Z. GOODFARB, Judge of the Superior Court of Maricopa County, Respondent,
Joseph K. BLEDSOE, Real Party in Interest.
Nos. CV=91=0122=SA, CV-91=0130&A.
Supreme Court of Arizona.
Dec. 19,1991.
Reconsideration Denied Feb. 19,1992.

Bicyclist who was severely injured when his bicycle struck "cable gate" across canal road brought suit against private corporation that operated and maintained irrigation canals. After partial summary judgment was entered in private corporation's favor, bicyclist brought special action petitions seeking review. The Supreme Court, Feldman, V.C.J., held that: (1) exceptional circumstances justified accepting jurisdiction of special action petitions to resolve questions raised by trial court's resolution of summary judgment motions; (2) canal road was not "agricultural land" within immunity provision of recreational use statue; (3) improved canal road located in urban community was not covered under recreational use statute's "other similar lands" provision so as to provide private corporation with immunity from liability for alleged negligence; and (4) *Salladay* doctrine did not give private corporation immunity for maintaining cable gate across canal road.

Order vacated, remanded.

Cameron, J., filed dissenting opinion.

1. Special actions challenging trial court's denial of summary judgment motions are appropriate only under exceptional circumstances. 17B A.R.S. Special Actions, Rules of Proc., Rule 1 (a).

2. Supreme Court would accept jurisdiction of special action petitions seeking relief from trial court's grant of partial summary judgment in case

which raised issues of statewide significance concerning the extent and continued viability of doctrine protecting owners and operators of irrigation canals from liability and interpretation of recreational use statute, which never had been considered by Supreme Court. 17B A.R.S. Special Actions, Rules of Proc., Rule l(a); A.R.S. V 33-1551.

3. Land falls into one of recreational use statute's enumerated types - agricultural, mining, range or forest - when, and only when, land is being used for such purpose; fact that land could be used, and has been used in the past, or could be used in future for such purpose does not transform it into agricultural, mining, range, or forest lands; also, land used for some purpose that merely suppports one of enumerated types is not land that is agricultural, mining, range, or forest land.

4. Roads designed to support canals and laterals maintained by private water association were not "agricultural" lands within immunity provisions of recreational use statue, inasmuch as association's roads simply supported canals that in turn provided water to agricultural lands, but were not land that was itself directly used for agricultural purposes; this land was not "premises" within meaning of recreational use statute. A.R.S. V 33-1551.

5. Improved canal road located in urban community was not covered under recreational use statute's catch-all "other similar lands" provision so as to provide immunity from negligence to private water association that maintained roads; land in question was man-made structure or road, small in size, transversing highly urbanized area, and thus was not the kind of natural land, true outdoors, or wilds that statute was designed to open to public enjoyment; road in question was graded and located in urban area within city limits. A.R.S. V 33-1551.

6. *Salladay* doctrine, limiting liability of operators of open flumes and canals carrying water, is inapplicable to canal roads if there is evidence that canal operator maintained road in manner that was unreasonably dangerous given foreseen and permitted use by public; doctrine is properly applied however, to claims based on theory that operation of unfenced canal and its equipment is per se act of negligence.

Cite as 277 Cal. Rptr. 887 (Cal. App. 2 Dist. 1991)
227 Cal.App.3d 134
Barbara BUCHAN, Plaintiff and Respondent,
v.
UNITED STATES CYCLING FEDERATION, INC.
Defendant and Appellant.
No. B037872.
Court of Appeal, Second District, Division 7.
Jan. 30, 1991.
Review Denied May 2, 1991.

Participant in bicycle race filed personal injury action against sponsor and sanctioning body for [head] injuries sustained in race while wearing hairnet helmet. The Superior Court, Los Angeles County, No. NWC94943, Irwin J. Nebron, J., entered judgment in favor of participant. Sponsor and sanctioning body appealed. The Court of Appeal, Fred Woods, J., held that release signed by participant barred action. [Cyclist's reasonable standard of care at elite-level bicycle racing was also discussed in opinion.]

Reversed and remanded with directions.

Johnson, J., filed a dissenting opinion.

1. Releases signed by participant in bicycle race, under which she expressly assumed all risks inherent in bicycle race, did not involve transactions affecting public interest and, thus, were not void as against public policy but, rather, effectively barred her negligence action for injuries sustained in race.

2. In interpreting written instrument, it is duty of appellate court to conduct de novo review and make determination in accordance with applicable principles of law.

382 So.2d 849 (Fla. Dist. Ct. App. 1980)
Charles CAPORALE and Karen Caporale, Appellants,
v.
RALEIGH INDUSTRIES OF AMERICA, INC.,
A Foreign Corporation, et. al., Appellees.
No. 799757.
District Court of Appeal of Florida, Third District.
April 22, 1980.

Products liability suit was brought against bicycle manufacturer-distributor and the retailer-assembler of the bicycle, seeking recovery for personal injuries. The Circuit Court, Dade County, Rhea P Grossman, J., directed verdicts for defendants at the close of evidence, and appeal was taken. The District Court of Appeal held that: (1) testimony of expert witness, coupled with plaintiff's statement, created jury question as to whether bicycle was improperly assembled and whether improper assembly caused accident, and (2) manufacturer-distributor would be liable for any injury caused by improper assembly of bicycle.

Reversed and remanded.

1. Testimony of expert witness that if bicycle had been properly assembled at time of purchase it would not have needed further adjustment nor would it have become dangerously loosened through normal use, coupled with plaintiff's statement that he had not tampered with assembly of bicycle since time of purchase, created jury question as to whether bicycle was improperly assembled by retailer and whether improper assembly caused accident.

2. Despite argument of bicycle manufacturer that authorized dealer was an independent contractor and that manufacturer was not responsible for dealer's action, manufacturer would be liable for any injury caused by improper assembly of bicycle since manufacturer instructed dealer in assembly of bicycle and implicitly made dealer's service its own.

3. Manufacturer cannot disclaim liability for injuries to ultimate purchaser of product where manufacturer knowingly relies on its dealer to put the product in a finished state.

Cite as 264 Cal. Rptr. 44 (Cal. App. 2 Dist. 1990)
215 Cal.App3d 934
NATIONAL AND INTERNATIONAL BROTHERHOOD
OF STREET RACERS, INC., Petitioner,
v.
The SUPERIOR COURT of Los Angeles County,
Respondent;
Richard CASTRO, Real Party in Interest.
CITY OF LOS ANGELES, Petitioner,
v.
The SUPERIOR COURT of Los Angeles County,
Respondent;
Richard CASTRO, Real Party in Interest.
Nos. B044644, B045173.
Court of Appeal, Second District, Division 2.
Nov. 16, 1989.
Review Denied Feb. 1, 1990.

Drag race organizer and landowner filed petitions for writ of mandate to require trial court to grant summary judgment to them in drag racer's personal injury action. The Court of Appeal, Compton, J., held that release protected organizer and landowner from liability for allegedly failing to assure presence of appropriate extrication equipment and properly trained rescue personnel.

Petitions granted.

1. Release signed by drag racer before race was unlimited in scope and, in unqualified terms, released all claims arising from the driver's participation in the race.

2. Release is sufficient if it is clear, unambiguous, and explicit and expresses agreement not to hold released party liable for negligence.

Cite as 581 So. 2d 1345 (Fla. App. 2 Dist. 1991)
Regala DENNIS, Appellant,
v.
CITY OF TAMPA, Appellee.
No. 90-01875.
District Court of Appeal of Florida, Second District.
May 29, 1991.
Certification Denied July 1 & 1991.

Park patron struck from rear by bicyclist brought personal injury action against city. The Circuit Court, Hillsborough County, J.C. Cheatwood, J., entered summary judgment in favor of city, and patron appealed. The District Court of Appeal, Altenbernd, J., held that: (1) city was not obligated to pay for supervision of park patrons engaged in typical recreation activities in order to prevent negligent conduct during low-use hours or in areas that were not expected to attract crowds, and (2) any duty owed by city to patron did not include specific standard of care under which patron could recover.

Affirmed.

557 A2d. 1068 (Pa. Super. Ct. 1989)
Robert HAVERSTOCK, a Minor, by his Parents and Natural Guardians,
Charles HAVERSTOCK and
Beverly Haverstock, and Charles and Beverly Haverstock, in Their Own Right
v.
T.I. RALEIGH (USA), Richard Goglia, t/a Oxford Valley Bicycle
Company, T.I. Raleigh, Ltd. and
The Huffy Corporation,
Appeal of T.I. RALEIGH (USA), T.I. Raleigh, Ltd.
and the Huffy Corporation.
Robert HAVERSTOCK, a Minor, by his Parents and Natural Guardians
Charles HAVERSTOCK and
Beverly Haverstock, and Charles and Beverly Haverstock, in Their Own Right
v.
T.I. RALEIGH (USA), Richard Goglia, t/a Oxford Valley Bicycle Company,
T.I. Raleigh, Ltd. and The Huffy Corporation.
Appeal of T.I. RALEIGH (USA), T.I. Raleigh, Ltd. and the Huffy Corporation.
Superior Court of Pennsylvania.
Argued Dec. 14,198s.
Filed March 31,1989.
Reargument Denied May 24,1989.

Parents brought suit seeking damages for injuries sustained by child when bicycle he was riding collapsed. The Court of Common Pleas, Civil Division, Philadelphia County, Nos. 8197, March Term, 1982 and 3803 December Term 1982, Prattis, J., found bicycle manufacturer and dealer liable, and defendants appealed. The Superior Court, Nos. 01816 and 02041 Philadelphia 1988, Cavanaugh, J., held that bicycle manufacturer was not, and could not be found vicariously liable as, a seller of a product made by another.

Reversed and remanded.

1. Having proceeded on theory of absolute liability, plaintiffs could not, after statute of limitations had expired, proceed on basis that defendants sold or distributed a chattel "manufactured by another."

2. Bicycle manufacturer was not, and could not be found vicariously liable as, a seller of a product made by another, based on its dealer's assembly of the bicycle and dealer's substitution of defective fork assembly.

ROBERT DAVID HIETT
v.
LAKE BARCROFT COMMUNITY ASSOCIATION, INC., ET. AL.
Record No. 911395
OPINION BY JUSTICE BARBARA MILANO KEENAN
June 5, 1992
From the Circuit Court of Fairfax County
Thomas J. Middleton, Judge

The primary issue in this appeal is whether a pre-injury release from liability for negligence is void as being against public policy. [We conclude here, based on Johnson, that the pre-injury release provision signed by Hiett is prohibited by public policy and, thus, it is void. Johnson, 86 Va. at 978, 11 SE. at 829.1 Robert D. Hiett sustained an injury which rendered him a quadriplegic while participating in the "Teflon Man Triathalon" (the triathlon) sponsored by the Lake Barcroft Community Association, Inc. (LABARCA). The injury occurred at the start of the swimming event when Hiett waded into Lake Barcroft to a point where the water reached his thighs, dove into the water, and struck his head on either the lake bottom or an object beneath the water surface.

Thomas M. Penland, Jr., a resident of Lake Barcroft, organized and directed the triathlon. He drafted the entry form which all participants were required to sign. The first sentence of the form provided:

> In consideration of this entry being accept(ed) to participate in the Lake Barcroft Teflon Man Triathlon I hereby, for myself, my heirs, and executors waive, release and forever discharge any and all rights and claims for damages which I may have or may here-after accrue to me against the organizers and sponsors and their representatives, successors, and assigns, for any and all injuries suffered by me in said event.

Evelyn Novins, a homeowner in the Lake Barcroft subdivision, asked Hiett to participate in the swimming portion of the triathlon. She and Heitt were both teachers at a school for learning-disabled children. Novins invited Hiett to participate as a member of one of two teams of fellow teachers she was organizing. During a break between classes, Novins presented Hiett with the entry form and he signed it.

Hiett alleged in his third amended motion for judgment that LABARCA, Penland, and Novins had failed to ensure that the lake was reasonably safe, properly supervise the swimming event, advise the participants of the risk of injury, and train them how to avoid such injuries. Hiett also alleged that Penland and Novins were agents of LABARCA and that Novin's failure to direct his attention to the release clause in the entry form constituted constructive fraud and misrepresentation.

In a preliminary ruling, the trial court held that, absent fraud, misrepresentation, duress, illiteracy, or the denial of an opportunity to read the form, the entry form was a valid contract and that the pre-injury release language in the contract released the defendants from liability for negligence. The trial court also ruled that such a release was prohibited as a matter of public policy only when it was included: (1) in a common carrier's contract of carriage; (2) in the contract of a public utility under a duty to furnish telephone service; or (3) as a condition of employment set forth in an employment contract.

Pursuant to an agreement between the parties, the trial court conducted an evidentiary hearing in which it determined that there was sufficient evidence to present to a jury on the issue of constructive fraud and misrepresentation. Additionally, the trial court ruled that as a matter of law Novins was not an agent of LABARCA, and it dismissed her from the case.

The remaining parties proceeded to trial solely on the issue whether there was constructive fraud and misrepresentation by the defendants such as would invalidate the waiver-release language in the entry form. After Hiett had rested his case, the trial court granted the defendants' motion to strike the evidence. This appeal followed.

Cite as 484 N.Y.S.2d 190 (App. Div. 1984)
106 A.D.2d 794
Michael S. HILLIS, an Infant, by Ann L. HANEY,
His Parent and Natural Guardian, et. al., Appellants,
v.
CITY OF ONEONTA, Respondent.
Supreme Court, Appellate Division, Third Department.
Dec. 20, 1984.

Action was brought against city on behalf of infant for personal injuries infant was alleged to have suffered as result of a fall from a city-owned bridge. Infant's father also brought derivative suit. The Supreme Court at Trial Term, Otsego County, Harlem, J., entered judgment on a jury verdict in favor of city, and appeals were taken. The Supreme Court, Appellate Division held that city's negligence in maintenance of bridge was proximate cause of 1 1-year-old bicyclist's fall off bridge to ground below and resultant injuries.

Reversed and remanded.

City's negligence in maintenance of bridge was proximate cause of 11 - year-old bicyclist's fall off bridge to ground below and resultant injuries.

Cite as 574 A.2d 934 (N.H. 1990)
Cain A. J. JOHNSTON, by his father and next friend James M. JOHNSTON, and
James M. Johnston, individually
v.
Frank LYNCH.
No. 88-058.
Supreme Court of New Hampshire.
April 30, 1990.

Suit was brought arising out of a bicycle-automobile collision. The Superior Court, Carroll County, McHugh J., entered judgment on a jury verdict in favor of defendant motorist, and bicyclist appealed. The Supreme Court, Thayer J., held that: (1) jury could reasonably have concluded that motorist was not negligent in operation of his automobile and therefore trial court did not err in failing to set aside verdict in favor of motorist on ground that it was against weight of the evidence; (2) trial court correctly excluded opinion testimony of investigating officer, and (3) trial court did not abuse its discretion in compelling plaintiffs to disclose identity of neurologist which they did not intend to call as a witness at trial since results of neurologist's examination of bicyclist were critical and irreplaceable evidence concerning bicyclist's condition during period of time when he would have suffered from postconcussion syndrome.

Affirmed.

1. Whether a jury verdict is against weight of the evidence is a separate issue from whether it is product of plain mistake, passion, partiality or corruption.

2. Jury could reasonably have concluded that motorist was not negligent in operation of his automobile, which collided with child bicyclist, and therefore trial court did not err in failing to set aside verdict in favor of motorist on ground that it was against weight of the evidence.

3. Verdict in favor of motorist in suit seeking damages for injuries sustained by child bicyclist in a collision with automobile was not product of plain mistake, passion, partiality or corruption.

4. Opinion testimony of investigating officer concerning fault and cause of automobile-bicycle collision was properly excluded; testimony would have involved mixed questions of law and the evidence on which his opinion rested

was available to jurors so that his testimony would not have assisted them in their search for the truth.

5. Even if opinion testimony bears directly on a main issue, evidence is admissible if it will help jury arrive at the truth.

6. Opinions of a police officer on fault and causation, which are mixed questions of law and fact, must be excluded.

7. Evidence established that motorist did not flee from scene of automobile-bicycle accident and therefore trial court did not abuse its discretion in excluding from jury evidence of circumstances surrounding motorist's leaving the scene.

In action arising from automobile-bicycle collision, trial court did not err in refusing portion of plaintiffs' requested instruction which involved a comment on the evidence and did not err in declining to give remainder of requested instruction on standard of reasonable care where court gave substance of that instruction in other language.

Cite as 425 N.W.2d 607 (Neb. 1988)
229 Neb. 114
**Lea Ann LAIRD, by John LAIRD, Her Next Friend and Father,
Appellant,**
v.
**Larry B. KOSTMAN, Appellee.
No. 86-418.
Supreme Court of Nebraska.
July 8, 1988.**

Bicyclist who was injured in collision with automobile brought negligence action against automobile's driver. The District Court, Dawson County, John I? Murphy, J., entered judgment in favor of defendant, and plaintiff appealed. The Supreme Court, Boslaugh, J., held that doctrine of last clear chance did not apply to action.

Affirmed.

Shamahn, J., concurred in result.

8. Doctrine of last clear chance did not apply to negligence action arising from collision between automobile and bicycle; bicyclist claimed that only two feet of bicycle extended into street beyond parked van when it was struck by automobile and that she saw automobile when it entered intersection 200 feet from point of impact, so that bicyclist's failure to move bicycle out of path of automobile was negligence which was active and contributing factor to accident.

**Cite as 250 Cal. Rptr. 299 (Cal. App. 2 Dist. 1988)
203 Cal.App.3d 589
Norman MADISON and YMCA of Metropolitan Los Angeles, Petitioners,
v.
The SUPERIOR COURT of the State of California for the County of Los Angeles, Respondent,
Sulejman and Maida SULEJMANAGIC, Real Parties in Interest
No. B033331.
Court of Appeal, Second District, Division 3.
Aug. 4,198s.
As Modified Sept. 1,198s.
Review Denied Oct. 13, 1988.**

Parents of scuba diving student brought action against organization and instructors for wrongful death of drowned student. The Superior Court, Los Angeles County, denied motion of instructor and organization for summary judgment. Organization petitioned for writ of mandate to direct trial court to vacate its denial of summary judgment. The Court of Appeal, Croskey, J., held that: (1) student's release of organization from liability for its negligence could not operate to limit parents' right to prosecute wrongful death action; (2) student assumed risk of injury by signing release and thereby relieved organization and instructors of any duty to him; and (3) release would have been defense in action brought by student and was defense to parents' wrongful death action.

Petition granted.

1. Scuba diving instructor, who was agent and employee of teaching organization and was acting within course and scope of agency in employment when student drowned, was entitled to legal protection of student's release of organization from all liability for its negligence, even though instructor was not specifically named in release.

2. Scuba diving student had no power or right to waive wrongful death action by his heirs, and, thus, student's express release of wrongful death action did not limit parents' right to prosecute action for wrongful death as result of drowning.

3. Decedent's express release covering negligence provides complete defense.

4. Release stating scuba diving student's intent to exempt and relieve organization from any liability for negligence expressly manifested intent to

assume risk and relieved organization and instructor of any duty in connection with dive that led to drowning.

5. For it to be valid and enforceable, written release exculpating tort-feasor from liability for future negligence or misconduct must be clear, unambiguous, and explicit in expressing intent of parties.

6. Scuba diving student's release of organization from liability for personal injury, property damage, or wrongful death caused by negligence was clear and free from ambiguity and was valid.

7. Absent public interest involvement, statute, which invalidates contract to exempt one from responsibility for willful injury or willful or negligent violation of law, does not invalidate contracts which seek to exempt one from liability for simple negligence or strict liability. West's Ann.Cal.Civ.Code V 1668.

8. Challenge to validity of scuba diving student's release of organization from liability for negligence involved no question of public interest and was not prohibited by statute, which invalidates contract to exempt one from responsibility for willful injury to person or willful and negligent violation of law.

West's Ann.Cal.Civ.Code V 1668.

9. Scuba diving student's release of organization from liability for negligence would have been complete defense to action brought by student and, therefore, was defense to his parents' action for wrongful death by drowning.

10. Scuba diving students' lack of knowledge about failure of instructor to follow buddy system rule and to stay with him at all times, that is, particular risk causing his death, did not invalidate student's release of organization and instructor for negligence and his express assumption of risk.

11. Plaintiff's specific knowledge of particular risk is required under doctrine of implied assumption of risk.

12. Scuba diving instructor's failure to follow buddy system, to supervise student at all times, and to prevent his drowning was reasonably related to object of purpose of student's release of organization and instructor from liability for negligence, and, thus, release was applicable to instructor's negligence.

13. If negligent act is reasonably related to subject of purpose for which release is given, it is reasonably foreseeable whether or not it is actually in contemplation of either party.

14. Claim that scuba diving instructor's failure to supervise student, to stay

with student at all times, and to prevent his drowning was not potential danger of scuba diving was irrelevant to determination whether student's release of instructors from liability for their negligence was applicable to instructor's failure to supervise.

342 So.2d 903 (Ala. 1977)
Jimmy Lee McWHORTER
v.
A.B. CLARK, as Administrator of the Estate of Henry C. Henderson, Deceased.
SC 1877.
Supreme Court of Alabama.
Feb. 18, 1977.

Parent filed suit against administrator of estate of deceased motorist for wrongful death of parent's eight-year-old son who lost his life in collision of automobile operated by motorist and bicycle being ridden by child. The Circuit Court, Houston County, Forrest L. Adams, J., entered judgement in favor of administrator, and parent appealed. The Supreme Court, Almon, J., held that police officer's testimony as to speed of motorist's automobile was improperly admitted and constituted prejudicial error; that under evidence, trial judge should not submit to jury issue of contributory negligence of deceased child; and that evidence as to absence of light on front of bike was admissible as relevant to initial negligence of motorist.

Reversed and remanded.

1. Automobile expert, not eyewitness to collision, can testify as to estimated speed of automobile predicated on distance the tires skidded or where dragged along the highway before impact, but not predicated on distance the tires skidded after impact.

2. In parent's action for wrongful death of his eight-year-old son who lost his life in collision of automobile operated by defendant motorist and bicycle being ridden by child, police officer's testimony as to speed of motorist's car, which testimony was based entirely on distance between point of impact and distance to car after it came to halt, was improperly admitted and constituted reversible error.

3. In parent's wrongful death action against administrator of estate of deceased motorist for wrongful death of parent's eight-year-old son who lost his life in collision of automobile operated by motorist and bicycle being ridden by the child, in absence of any evidence tending to show that deceased child possessed maturity and sensibility of 14-year-old, judge erred in submitting question of child's contributory negligence.

4. Regardless of merits of appellee's contentions, appellate court will not review questions not decided by trial court.

5. In parent's action against administrator of estate of deceased for wrongful death of parent's eight-year-old son who lost his life in collision of automobile operated by deceased motorist and bicycle being ridden by child, evidence as to absence of light on front of bike was admissible as relevant only to initial negligence of driver of automobile.

Cite as 571 N.E. 2d 1056 (Ill. App. 4 Dist. 1991)
213 Ill. App. 3d 217
MOORE
v.
SWOBODA
Nos. 4-89-0957, 4-90-0044.
Appellate Court of Illinois, Fourth District.
April 23, 1991.

Father of minor dirt bike rider struck and killed by truck, as special administrator, brought survival action against truck driver and driver's employer for rider's personal injuries and death, also seeking damages for wrongful death and, in his individual capacity, for medical and funeral expenses. The Circuit Court, Vermilion County, Rita B. Garman, J., entered judgment on jury verdicts awarding damages for medical and funeral expenses and for rider's pain and suffering, but allowed defendants' motion for new trial on wrongful death counts. Granting separate petitions for leave to appeal by plaintiff and defendants, the Appellate Court, Steigmann, J., held that: (1) lack of proper instruction to jury that contributory negligence was complete bar to recovery of damages for pecuniary injuries on wrongful death claim did not require new trial, in light of finding that both of rider's parents were negligent; (2) evidence supported conclusion that truck driver negligently breached duty of ordinary care by not taking appropriate steps to avoid collision; but (3) jury instruction that both driver and rider were to be held to standard of ordinary care of adult should have been given; and (4) jury should not have been instructed that an issue was whether driver kept proper lookout for "children" near highway.

Reversed and remanded with directions; reversed and remanded for new trial.

7. Evidence supported conclusion that truck driver negligently breached duty of ordinary care by not taking appropriate steps to avoid collision with dirt bike; jury could reasonably have concluded that had driver braked when he first saw dirt bike, collision would not have occurred or impact would have been lessened and that sounding truck's horn could have prevented accident, and jury could have discounted truck driver's testimony that he was driving only 45 to 50 miles per hour given expert testimony suggesting greater speed.

15. Minors operating motor vehicles and minibikes on public streets and roads are subject to same standard of care as adults.

17. Jury should not have been instructed that an issue was whether driver, of truck which struck minor dirt bike rider, kept proper lookout for "children"

near highway; such instruction may have confused jurors, as rider was subject to adult standard of care.

26. Absent evidence that head injury caused death of minor dirt bike rider struck by truck, evidence as to whether rider was wearing helmet at time of accident was not relevant to issue of what caused his death and was properly excluded in negligence action arising from accident.

27. Generally, evidence of failure to wear motorcycle helmet or other protective devices, such as seat belts, is inadmissible to establish contributory negligence.

Cite as 476 N.E. 2d 1378 (1985)
132 Ill.App.3d 90 87 Ill.Dec.386
Russell RALSTON, Administrator of the Estate of Romney Ralston, Deceased, Plaintiff-Appellee,
v.
Hal W. PLOGGER and Houlette and Seaton, Inc., Defendants-Appellants.
No. 4-84-0628.
Appellate Court of Illinois, Fourth District.
April 10, 1985.

Administrator of minor decedent's estate brought wrongful death action against truck driver and his employer. The jury returned a verdict in favor of defendants, but the Circuit Court, County of Macoupin, John W. Russell, J., granted administrator's motion for new trial. Upon granting defendants leave to appeal, the Appellate Court, McCullough, J., held that: (1) any error regarding damage instruction was not so pervasive or prejudicial as to create a likelihood that it may have affected jury's decision on issue of liability; (2) trial court properly instructed jury on issue of contributory negligence; (3) trial court's instructions were sufficiently clear so as not to mislead the jury, they fairly and accurately stated the law, and they did not unduly emphasize any particular matter, and (4) verdict was not against manifest weight of evidence.

Reversed and remanded.

Trapp, J., filed opinion concurring in part and dissenting in part.

12. A witness whose knowledge is based primarily on practical experience is no less an expert than one who possesses particular scientific or academic knowledge.

13. Question of witness' knowledge or degree of his experience goes to weight accorded his testimony not to his competency.

14. Reconstruction testimony may not be used as substitute for eyewitness testimony; whether such expert testimony may be used in addition to eyewitness testimony is determined by whether it is necessary to rely on application of principles of science beyond ken of ordinary juror; if matter is not beyond ken of average juror, then it is inadmissible.

16. Admission of deputy's testimony concerning point of impact did not cause prejudicial error in wrongful death action brought by administrator of

minor decedent's estate against truck driver and his employer.

18. Verdict in favor of defendants in wrongful death action involving nine-year-old decedent was not against manifest weight of evidence.not

Cite as 515 N.E. 2d 298 (Ill. App. 1 Dist. 1987)
161 Ill. App. 3d 742
SCHAFFNER
v.
CHICAGO & NORTH WESTERN TR.
No. 884200.
Appellate Court of Illinois, First District, Third Division.
September 23, 1987.

Bicyclist who sustained serious injuries when front wheel of his bicycle disengaged as he rode across railroad tracks brought suit against railroad and bicycle manufacturer. The Circuit Court, Cook County, Anthony J. Bosco, J., entered judgment on a jury verdict of $8,235,000 against railroad, and entered judgment in favor of bicycle manufacturer. Railroad appealed from both judgments. Plaintiffs appealed from judgment in favor of bicycle manufacturer. The Appellate Court, McNamara, P.J., held that: (1) railroad failed to show that it was prejudiced by evidence of post-accident replacement of railroad crossing; (2) verdict against railroad was not excessive; and (3) there was sufficient evidence for jury to conclude that bicycle's design was not unreasonably dangerous and that when accident occurred a safer, feasible alternative to design of quick release mechanism was not available to manufacturer.

Judgments affirmed.

19. Evidence that many others had used bicycle with quick release front wheel without incident was admissible in product liability suit arising out of accident which occurred when front wheel disengaged when bicyclist was going over railroad crossing; evidence was relevant to issue whether design of bicycle and quick release mechanism was defectively dangerous.

22. Instruction on feasibility of alternative designs that manufacturer has no duty to manufacture different design if different design is not feasible, and that feasibility includes not only elements of economy, effectiveness and practicality, but also technological possibilities under the state of the manufacturing art at the time the product was manufactured, did not improperly place burden of proof on plaintiffs in products liability suit.

26. Evidence was sufficient in product liability suit arising from accident which occurred when quick release front wheel of bicycle disengaged as bicycle was being ridden over railroad crossing for jury to conclude that bicycle's design was not unreasonably dangerous and that at time accident occurred a safe, feasible alternative was not available to manufacturer, there was expert

testimony that bicycle design was safe because quick release mechanism, when properly secured, would remain closed even if bicycle was ridden over a bumpy road.

Corboy & Demetrio, Chicago (Philip H. Corboy, Bruce Robert Pfaff and David A. Novoselsky, of counsel), for the Schaffners.

Baker & McKenzie, Chicago (Francis D. Morrissey, Michael K. Murtaugh, Marie A. Monahan and John A. Krivicich, of counsel), for Schwinn Bicycle.

475 So.2d 526 (Ala. 1985)
Willie Jean SMITH, as mother of Willie Earl Conner, a deceased minor
v.
Ralph BRADFORD.
No. 83-1253.
Supreme Court of Alabama.
Aug. 23, 1985.

Action was brought against state trooper for wrongful death of minor, alleging negligence and wanton misconduct, arising out of accident wherein child on bicycle was struck by state trooper's vehicle while trooper was in pursuit of motorist. The Circuit Court, Tuscaloosa County, Claude Harris, Jr., J., entered judgment upon jury verdict for state trooper, and parent of deceased child appealed. The Supreme Court, Maddox, J., held that: (1) evidence was sufficient to carry case to the jury on issue of state trooper's wantonness, and (2) evidence was sufficient to carry the case to the jury on issue of whether child who was 13 years, 7 months old possessed maturity and sensibility of ordinary 14.year-old so that child was legally capable of contributory negligence.

Affirmed in part; reversed in part, and remanded.

1. Upon appeal from trial court's grant of directed verdict in favor of defendant on wantonness count in wrongful death action, standard of review is not whether there was genuine issue of material fact, but whether there was sufficient evidence to prevent directed verdict on wantonness count.

2. Question of whether there was proof of wantonness must be determined by facts and circumstances of each case.

3. Upon appeal from trial court's directed verdict for defendant on wantonness count in wrongful death action, Supreme Court must consider evidence most favorable to plaintiff.

4. In action against state trooper for wrongful death of minor, arising out of incident wherein child on bike was killed when hit by state trooper's vehicle while vehicle was in pursuit of motorist, evidence was sufficient to carry case to the jury on issue of state trooper's wantonness, where state trooper acknowledged that he did not turn on his emergency light or siren, that he was travelling in excess of 80 miles per hour and believed himself in hazardous and perilous situation, and that he started releasing his brakes under the assumption that he had passed bicyclist.

5. Child between the ages of 7 and 14 is prima facie incapable of contribu-

tory negligence, but may be shown to be capable of contributory negligence by proving that he possessed that discretion, intelligence, and sensitivity to danger which ordinary 14-year-old possesses.

6. Evidence was sufficient to carry case to the jury on issue of contributory negligence of child who as 13 years, 7 months old, in action against state trooper for wrongful death of minor, arising out of accident wherein child on bike was struck by state trooper's vehicle while trooper was in pursuit of motorist, where there was at least scintilla of evidence that child possessed maturity and sensibility of ordinary 14-year-olds.

Cite as 459 N.W. 2d 151 (Minn. App. 1990)
George D. SWANLUND and Lu Jean Swanlund,
Appellants,

v.

SHIMANO INDUSTRIAL CORPORATION, LTD., a Japanese corporation,
and Shimano American Corporation, a United States corporation,
Respondents.
No. CO-90-243.
Court of Appeals of Minnesota.
August 14,1990.
Review Denied October 5,1990.

Bicyclist filed products liability action against manufacturer and distributor of quick-release bicycle hub. The District Court, Hennepin County, Andrew W. Danielson, J., denied the bicyclist's motion to amend the complaint to plead punitive sustain compensatory damage award, but not punitive damages. Bicyclist petitioned for discretionary review. The Court of Appeals, Short, J., held that: (1) on a motion to amend to plead punitive damages, a party must offer evidence which, if unrebutted, would be clear and convincing evidence of a willful indifference to the rights or safety of others; (2) the appropriate standard in a pretrial appeal from a refusal to allow the amendment is that applied to a directed verdict; (3) the bicyclist's evidence showed at most negligence, which was insufficient to establish willful indifference; and (4) discretionary review of orders granting or denying motions to add claims for punitive damages will not be allowed absent exceptional circumstances.

Affirmed.

Lansing, J., concurred specially with opinion.

8. Single incident of defectively manufactured quick-release bicycle hub that separated despite being correctly attached to bicycle fork was insufficient to support finding of willful indifference to safety of others, and thus, trial court did not abuse its discretion in denying bicyclist's motion for leave to amend complaint to assert claim for punitive damages.

M.S.A. 549.191.

9. Manufacturer's failure to assure that bicycle assembler incorporated wheel retention device upon assembly of bicycle with quick-release hub was at most negligence, which was insufficient to establish willful indifference, and thus, trial court did not abuse its discretion in refusing to allow bicyclist to

amend products liability complaint to assert claim for punitive damages. M.S.A. 549.19 1.

10. Manufacturer's failure to advise customers and public of risk of wheel separation on bicycles with quick-release hubs was at most negligent failure to warn, which could sustain compensatory damage award, but not punitive damage award. M.S.A. 549.191.s;

WALKER v. CITY OF SCOTTSDALE
Cite as 786 P2d 1057 (Ariz. App. 1989)
163 Ariz. 206
Myrna N. WALKER, Plaintiff-Appellant,
CITY OF SCOTTSDALE, a political subdivision of the State of Arizona, and the McCormick Range Property Owners' Association, Defendants-Appellees.
No. 1 CA-CV 88-256.
Court of Appeals of Arizona, Division 1, Department D.
November 24, 1989.
Review Denied February 26, 1990.

Bicyclist who was injured in fall on maintained bike path running through greenbelt area of urban residential neighborhood brought suit against owner's association and city, which owned easement on association property for bicycle path. The Superior Court, Maricopa County, Cause No. CV 87-08614, Robert W. Pickrell, J., entered summary judgment in favor of city and owner's association, finding that they had immunity from suit under recreational user's statute. Bicyclist appealed. The Court of Appeals, Nelson, J., held that greenbelt area in urban, residential neighborhood did not fall within definition of "premises" contained in recreational user's statute, and thus defendants did not enjoy limited immunity from liability under statute.

Reversed and remanded.

1. Urban greenbelt area located within planned, residential community and owned by homeowner's association did not fall within definition of "premises" in recreational users statute so as to afford association and city limited immunity from liability in suit filed by bicyclist alleging that injuries she suffered in fall from bike while riding bicycle path easement through greenbelt were caused by their negligence. A.R.S. 33-1551. See publication Words and Phrases for other judicial construction and definitions.

2. Language of statute is most reliable evidence of its intent.

5. Phrase "any other similar lands" in definition of "premises" contained in recreational user statute, which provided limited immunity to owners, lessees or occupants of "agricultural, range, mining, or forest lands and other similar lands," could not be construed as applying to any land upon which, by agreement, any of statue's stated recreational activities could be performed, inasmuch as remainder of definition of premises, "agricultural, range, mining or

forest lands" would then become surplusage. A.R.S. 33-1551.

7. Recreational users statue did not limit liability of owner's association and city with respect to maintained bike path running through greenbelt area of urban-residential neighborhood in suit brought by bicyclist injured on bike path.

A.R.S. 334551.

Dake, Hathaway & Goering, PA. by Tersca W. Goering and T. Gale Dake, Phoenix, for plaintiff-appellant.

Engineering Appendix C

Federal Regulations

Code of Federal Regulations

Commercial Practices 16
PART 1000 TO END
Revised as of January 1, 1994
CONTAINING A CODIFICATION OF DOCUMENTS OF GENERAL APPLICABILITY AND FUTURE EFFECT
AS OF JANUARY 1, 1994
With Ancillaries Published by the Office of the Federal Register National Archives and Records Administration as a Special Edition of the Federal Register

Part 1512, requirements for Bicycles

Subpart A-Regulations
Sec
1512.1 Scope.
1512.2 Definitions.
15 12.3 Requirements in general.
15 12.4 Mechanical requirements.
15 12.5 Requirements for braking system.
15 12.6 Requirements for steering system.
15 12.7 Requirements for pedals.
15 12.8 Requirements for drive chain.
15 12.9 Requirements for protective guards.
15 12.10 Requirements for tires.
15 12.11 Requirements for wheels.
15 12.12 Requirements for wheel hubs.
15 12.13 Requirements for front fork.
15 12.14 Requirements for fork and frame assembly.
1512.15 Requirements for seat.
15 12.16 Requirements for reflectors.
15 12.17 Other requirements.
15 12.18 Tests and test procedures.
15 12.19 Instructions and labeling.
15 12.20 Separability.

Subpart B-Policies and interpretations
15 12.50 Affirmative labeling statement.
FIGURES 1-8
TABLES 1-4
AUTHORITY Sets. 2(f)(l)(D), (q)(l)(A), (s), 3(e)(l), 74 Stat. 372,374,375, as amended, 80 Stat. 1304-05, 83 Stat. 187-89 (15 U.S.C. 1261, 1262).
SOURCE: 43 FR 60034, Dec. 22, 1978, unless otherwise noted.

Subpart A-Regulations

§ 1512.1 Scope.

This part sets forth the requirements for a bicycle as defined in § 1512.2(a) (except a bicycle that is a track bicycle or a one-of-a-kind bicycle as defined in § 1512-2 (d) and (e)) which is not a banned article under § 1500.18(a)(12) of this chapter.

§ 1512.2 Definitions.

For the purposes of this part:

(a) Bicycle means a two-wheeled vehicle having a rear drive wheel that is solely human-powered.

(b) *Sidewalk bicycle* means a bicycle with a seat height of no more than 635 mm (25.0 in); the seat height is measured with the seat adjusted to its highest position.

(c) Seat height means the dimension from the point on the seat surface intersected by the seat post center line (or the center of the seating area if no seat post exists) and the ground plane, as measured with the wheels aligned and in a plane normal to the ground plane.

(d) *Track bicycle* means a bicycle designed and intended for sale as a competitive machine having tubular tires, single crank-to-wheel ratio, and no freewheeling feature between the rear wheel and the crank.

(e) One-of-a-kind bicycle means a bicycle that is uniquely constructed to the order of an individual consumer other than by assembly of stock or production parts.

(f) Normal riding position means that the rider is seated on the bicycle with both feet on the pedals and both hands on the handlegrips (and in a position that allows operation of handbrake levers if so equipped); the seat and handlebars may be adjusted to positions judged by the rider to be comfortable.

§ 1512.3 Requirements in general.

Any bicycle subject to the regulations in this part shall meet the requirements of this part in the condition to which it is offered for sale to consumers; any bicycle offered for sale to consumers in disassembled or partially assembled condition shall meet these requirements after assembly according to the manufacturer's instructions. For the purpose of compliance with this part, where the metric and English units are not equal due to the conversion process the less stringent requirement will prevail.

δ 1612.4 Mechanical requirements.

(a) *Assembly*. Bicycles shall be manufactured such that mechanical skills required of the consumer for assembly shall not exceed those possessed by an adult of normal intelligence and ability.

(b) *Sharp edges*. There shall be no unfinished sheared metal edges or other sharp parts on bicycles that are, or may be, exposed to hands or legs; sheared metal edges that are not rolled shall be finished so as to remove any feathering of edges, or any burrs of spurs caused during the shearing process.

(c) Integrity. There shall be no visible fracture of the frame or of any steering, wheel, pedal, crank, or brake system component resulting from testing in accordance with: The handbrake loading and performance test, δ 1512.18(d); the foot brake force and performance test, δ 1512.18(e); and the road test, δ 1512.18(p) (or the sidewalk bicycle proof test, δ 1512.18(q)).

(d) *Attachment hardware*. All screws, bolts, or nuts used to attach or secure components shall not fracture, loosen, or otherwise fail their intended function during the tests required in this part. ALL threaded hardware shall be of sufficient quality to allow adjustments and maintenance. Recommended quality thread form is specified in Handbook H28, "Screw Thread Standards for Federal Service,"[1] issued by the National Bureau of Standards, Department Of Commerce; recommended mechanical properties are specified in ISO Recommendation R898, "Mechanical Properties of Fasteners," and in ISO Recommendations 68, 262, and 263, "General Purpose screw Threads."[2]

(e)-(f) [Reserved]

(g) Excluded area. There shall be no protrusions located within the area bounded by (1) a line 89mm (3-1/2 in) to the rear of and parallel to the handlebar stem; (2) a line tangent to the front tip of the seat and intersecting the seat mast at the top rear stay; (3) the top surface of the top tube; and (4) a line con-

necting the front of the seat (when adjusted to its highest position) to the junction where the handlebar is attached to the handlebar stem. The top tube on a female bicycle model shall be the seat mast and the down tube or tubes that are nearest the rider in the normal riding position. Control cables no greater than 6.4 mm (1/4 in) in diameter and cable clamps made from material not thicker than 4.8 mm (3/16 in) may be attached to the top tube.

(h) [Reserved]

(i) *Control cable ends.* Ends of all control cables shall be provided with protective caps or otherwise treated to prevent unraveling. Protective caps shall be tested in accordance with the protective cap and end-mounted devices test, § 1512.18(c), and shall withstand a pull of 8.9 N (2.0 lbf).

(i) *Control cable abrasion.* Control cables shall not abrade over fixed parts and shall enter and exit cable sheaths in a direction in line with the sheath entrance and exit so as to prevent abrading.

§ 1512.5 Requirement for braking system.

(a) *Braking system.* Bicycles shall be equipped with front- and rear-wheel brakes or rear-wheel brakes only.

(b) *Handbrakes.* Handbrakes shall be tested at least ten times by applying a force sufficient to cause the handlever to contact the handlebar, or a maximum Of 445 N (100 lbf), in accordance with the loading test, § 1512.18(d)(2). and shall be rocked back and forth with the weight Of a 68.1 kg (150 lb) rider on the seat with the same handbrake force applied in accordance with the rocking test, § 1512.18(d)(2)(iii); there shall be no visible fractures, failures, movement of clamps, or misalignment of brake components.

(1) *Stopping distance.* A bicycle equipped with only handbrakes shall be tested for stopping distance by a rider of at least 68.1 kg (150 lb) weight in accordance with the performance test, § 1512.18(d)(2)(v) and (vi), and shall have a stopping distance of no greater than 4.67 m (15 ft) from the actual test speed as determined by the equivalent ground speed specified in § 1512.18(d)(2)(vi).

(2) *Hand lever access.* Hand lever mechanisms shall be located on the handlebars in a position that is readily accessible to the rider when in a normal riding position.

(3) *Grip dimension.* The grip dimension (maximum outside dimension between the brake hand lever and the handlebars in the plane containing the centerlines of the handgrip and the hand brake lever) shall not exceed 89 mm (3-1/2 in) at any point between the pivot point of the lever and lever midpoint; the grip dimension for sidewalk bicycles shall not exceed 76 mm (3 in). The grip

dimension may increase toward the open end of the lever but shall not increase by more than 12.7 mm (1/2 in) except for the last 12.7 mm (1/2 in) of the lever. (See figure 5 of this part 1512.)

(4) Attachment. Brake assemblies shall be securely attached to the frame by means of fasteners with locking devices such as a lock washer, locknut, or equivalent and shall not loosen during the rocking test, δ 1512. M(d)- (2)(iii). The cable anchor bolt shall not cut any of the cable strands.

(5) *Operating Force.* A force of less than 44.5 N (10 lbf) shall cause the brake pads to contact the braking surface of the wheel when applied to the handlever at a point 25 mm (1.0 in) from the open end of the handlever.

(6) Pad and pad holders. Caliper brake pad shall be replaceable and adjustable to engage the braking surface without contacting the tire or spokes and the pad holders shall be securely attached to the caliper assembly. The brake pad material shall be retained in its holder without movement when the bicycle is loaded with a rider of at least 68.1 kg (150 lb) weight and is rocked forward and backward as specified in the rocking test, δ 1512XS(d)(2)(iii).

(7) [Reserved]

(8) Hand lever location. The rear brake shall be actuated by a control located on the right handlebar and the front brake shall be actuated by a control located on the left handlebar. The left-hand/right-hand locations may be reversed in accordance with an individual customer order. If a single hand lever is used to actuate both front and rear brakes, it shall meet all applicable requirements for hand levers and shall be located on either the right or left handlebar in accordance with the customer's preference.

(9) Hand lever extensions. Bicycles equipped with hand lever extensions shall be tested with the extension levers in place and the hand lever extensions shall also be considered to be hand levers.

(c) Footbrakes. All footbrakes shall be tested in accordance with the force test, δ 1512.18(e)(2), and the measured braking force shall not be less than 178 N (40 lbf) for an applied pedal force of 310 N (70 lbf).

(1) *Stopping distance.* Bicycles equipped with footbrakes (except sidewalk bicycles) shall be tested in accordance with the performance test δ 15 12.18(e)(3), by a rider of at least 68.1 kg (150 lb) weight and shall have a stopping distance of no greater than 4.57 m (15 ft) from an actual test speed of at least 16 km/h (10 mph). If the bicycle has a footbrake only and the equivalent

groundspeed of the bicycle is in excess of 24 km/h (15 mph} (in its highest gear ratio at a pedal crank rate of 60 revolutions per minute),[3] the stopping distance shall be 4.57 m (15 ft) from an actual test speed of 24 km/h (16 mph) or greater.

(2) Operating force. Footbrakes shall be actuated by a force applied to the pedal in a direction opposite to that of the drive force, except where brakes are separate from the drive pedals and the applied force is in the same direction as the drive force.

(3) Crank differential. The differential between the drive and brake positions of the crank shall be not more than 60 with the crank held against each position under a torque of no less than 13.6 N-m (10 ft-lb).

(4) Independent operation. The brake mechanism shall function independently of any drive-gear positions or adjustments.

(d) *Footbrakes and handbrakes in combination.* Bicycles equipped with footbrakes and handbrakes shall meet all the requirements for footbrakes in δ 1512.5(c), including the tests specified. In addition, if the equivalent ground speed of the bicycle is 24 km/h (15 mph) or greater (in its highest gear ratio at a pedal crank rate of 60 revolutions per minute),3 the actual test speed specified in δ 1512.18(e)(3) shall be increased to 24 km/h (15 mph) and both braking systems may be actuated to achieve the required stopping distance of 4.57 m (15 ft).

(e) *Sidewalk bicycles.* (1) Sidewalk bicycles shall not have handbrakes only.

(2) Sidewalk bicycles with a seat height of 560 mm (22 in) or greater (with seat height adjusted to its lowest position) shall be equipped with a footbrake meeting all the footbrake requirements of δ 1512.5(c), including the specified tests except that the braking force transmitted to the rear wheel shall be in accordance with the sidewalk bicycle footbrake force tests, δ 1512.18(f).

(3) Sidewalk bicycles with a seat height less than 560 mm (22 in) (with seat height adjusted to its lowest position) and not equipped with a brake shall not have a freewheel feature. Such sidewalk bicycles equipped with a footbrake shall be tested for brake force in accordance with the sidewalk bicycle footbrake force test, δ 51512.18(f). Such sidewalk bicycles not equipped with brakes shall be identified with a permanent label clearly visible from a distance of 3.1 m (10 ft) in daylight conditions and promotional display material and shipping cartons shall prominently display the words "No Brakes."

δ 1512.6 Requirements for steering systems.

(a) *Handlebar stem insertion mark.* The handlebar stem shall contain a permanent ring or mark which clearly indicates the minimum insertion depth of the handlebar stem into the fork assembly. The insertion mark shall not affect the structural integrity of the stem and shall not be less than 2-1/2 times the stem diameter from the lowest point of the stem. The stem strength shall be maintained for at least a length of one shaft diameter below the mark.

(b) *Handlebar stem strength.* The handlebar stem shall be tested for strength in accordance with the handlebar stem test, δ 1512.18(g), and shall withstand a force of 2000 N (450 lbf) for bicycles and 1000 N (225 lbf) for sidewalk bicycles.

(c) *Handlebar.* Handlebars shall allow comfortable and safe control of the bicycle. Handlebar ends shall be symmetrically located with respect to the longitudinal axis of the bicycle and no more than 406 mm (16 in) above the seat surface when the seat is in its lowest position and the handlebar ends are in their highest position.

(d) *Handlebar ends.* The ends of the handlebars shall be capped or otherwise covered. Handgrips, end plugs, control shifters, or other end-mounted devices shall be secure against a removal force of no less than 66.8 N (15 lbf) in accordance with the protective cap and endmounted devices test, δ 15 12.18(c).

(e) *Handlebar and clamps.* The handlebar and clamps shall be tested in accordance with the handlebar test, δ 1512.18(h). Directions for assembly of the bicycle required in the instruction manual by δ 1512.19(a)(2) shall include an explicit warning about the danger of damaging the stem-to-fork assembly and the risk of injury to the rider that can result from over tightening the stem bolt or other clamping device. The directions for assembly shall also contain a simple, clear, and precise statement of the procedure to be followed to avoid damaging the stem-to-fork assembly when tightening the stem bolt or other clamping device.

6 1515.7 Requirements for pedals.

(a) *Construction.* Pedals shall have right-hand/left-hand symmetry. The tread surface shall be present on both top and bottom surfaces of the pedal except that if the pedal has a definite preferred portion, the tread surface need only be on the surface presented to the rider's foot.

(b) *Toe* clips. Pedals intended to be used only with toe clips shall have toe lips securely attached to them and need not have tread surfaces. Pedals designed or optional use of toe clips shall have tread surfaces.

(c) Pedal reflectors. Pedals for bicycles other than sidewalk bicycles shall have reflectors in accordance with δ 1512.16(e). Pedals for sidewalk bicycles are not required to have reflectors.

δ 1512.8 Requirements for drive chain.

The drive chain shall operate over the sprockets without catching or binding. The tensile strength of the drive chain shall be no less than 8010 N (1,800 lbf) or 6230 N (1,400 lbf) for sidewalk bicycles.

δ 1512.9 Requirements for protective guards.

(a) *Chain guard.* Bicycles having a single front sprocket and a single rear

(iv) Shape the cup brush by hand to the specified 0.5 (approx. 13mm) diameter. Any stray wire bristles projecting more than 1/32 in. (approx. 1 mm) beyond the tip of the bulk of the bristles should be clipped off. Adjust the position of the brush so that its axis is centered over the mid-point in the width of the retroreflective material.

(v) Adjust the rotational velocity of the bicycle wheel to obtain a linear velocity of 0.23 m/set (9 in./sec) measured at the mid-point in the width of the retroreflective material. Adjust the force to obtain a force normal to the surface under the brush of 2 N (0.45 lbf).

(vi) Apply the abrading brush to the retroreflective material on the wheel rim, and continue the test for 1000 complete revolutions of the bicycle wheel. [43 FR 60034, Dec. 22,1978, as amended at 45 FR 82628, Dec. 16,1980,46 FR 3204, Jan. 14, 19811

δ 1512.19 Instructions and labeling.

A bicycle shall have an instruction manual attached to its frame or included with the packaged unit.

(a) The instruction manual shall include at least the following:

(1) Operations and safety instructions describing operation of the brakes and gears, cautions concerning wet weather and night-time operation, and a guide for safe on-and-off road operation.

(2) Assembly instructions for accomplishing complete and proper assembly.

(3) Maintenance instructions for proper maintenance of brakes, control cables, bearing adjustments, wheel adjustments, lubrication, reflectors, tires and handlebar and seat adjustments; should the manufacturer determine that such maintenance is beyond the capability of the consumer, specifics regarding

locations where such maintenance service can be obtained shall be included.

(b) A bicycle less than fully assembled and fully adjusted shall have clearly displayed on any promotional display material and on the outside surface of the shipping carton the following: (1) A list of tools necessary to properly accomplish assembly and adjustment. (2) a drawing illustrating the minimum leg-length dimension of a rider and a method of measurement of this dimension.

16 CFR Ch. 11 (l-1-94 Edition)-

(c) The minimum leg-length dimension shall be readily understandable and shall be based on allowing no less than one inch of clearance between (1) the top tube of the bicycle and the ground plane and (2) the crotch measurement of the rider. A girl's style frame shall be specified in the same way using a corresponding boys' model as a basis.

(d) Every bicycle subject to the requirements of this part 1512 and introduced into interstate commerce on or after May 11, 1976 through May 11,1978, shall be labeled with the statement "Meets U.S. Consumer Product Safety Commission Regulations for Bicycles ."

(1) Every such bicycle, displayed or offered for sale to consumers in a fully assembled condition, shall bear a label (such as a hang tag) at least 6.4 cm (2.5 in.) by 17.8 cm (7 in.) setting forth the required labeling statement legibly and conspicuously in capital letters at least 0.6 cm (0.25 in.) high. No other words or symbols may appear on the label. (See also 8 1512.50.)

(2) The required labeling statement shall appear legibly and conspicuously in capital letters at least 1.3 cm (0.5 in.) high on the retail carton of every such bicycle offered for sale to consumers in an unassembled or partially assembled condition.

(e) Every bicycle subject to the requirements of this part 15 12 shall bear a marking or label that is securely affixed on or to the frame of the bicycle in such a manner that the marking or label cannot be removed without being defaced or destroyed. The marking or label shall identify the name of the manufacturer or private labeler and shall also bear some form of marking from which the manufacturer can identify the month and year of manufacture or from which the private labeler can identify the manufacturer and the month and year of manufacture. For purposes of this paragraph, the term *manufacture* means the completion by the manufacturer of a bicycle of those construction or assembly operations that are performed by the manufacturer before the bicycle is shipped from the manufacturer's place of production for sale to distributors, retailers, or consumers.

Consumer Product Safety Commission

§ 1512.20 Separability.

If any section or portion thereof of this part 1512 or its application to any person or circumstance is held invalid, the remainder of the section(s) and its (their) application to other persons or circumstances is not thereby affected.

Subpart B-Policies and Interpretations

§ 1512.50 Affirmative labeling statement.

(a) Section 1512.19(d) requires every bicycle subject to the requirements of this part 1512 introduced into interstate commerce on or after May 11, 1976 through May 11, 1978, to be labeled with the statement "Meets U.S. Consumer Product Safety Commission Regulations for Bicycles." In accordance with § 1512.19(d)(1), the label on each assembled bicycle, which may consist of a hang tag, is required to be at least 6.4 cm. (2.5 in.) by 17.8 cm. (7 in.) with the labeling statement in capital letters at least 0.6 cm. (0.25 in.) high.

(b) Because of variances in the manufacture of hang tags, a finished tag, ordered to the specifications of § 1512.19(d)(1), may be slightly smaller than the minimum specifications. However, the Commission finds that hang tags with either length or width dimensions (or both) of no more than 0.32 cm. (1/8 in.) less than the prescribed requirements adequately provide the requisite degree of conspicuousness to consumers.

(c) Therefore, the Commission will consider bicycles otherwise in compliance with the provisions of part 1512 to be in compliance with the requirements as to length and width of hang tags used to comply with labeling requirements under § 1512.19(d)(1) for purposes of enforcement if:

(1) The hang tag is correctly labeled with the required statement under § 1512.19(d), and

(2) The hang tag meets all of the labeling conspicuousness, legibility, and type size requirements of § 1512.19(d)(l), and

(3) It can be documented that the hang tag was ordered to the correct specifications but, due to a manufacturing variance, is no more than 0.32cm. (1/8 in.) smaller in either or both of its linear dimensions than the requirements of § 1512.19(d)(l).

(Sec. 10(a), 74 Stat. 378;15 U.S.C. 1269(a))

1. Copies may be obtained from: Superintendent of Documents U.S. Government Printing Office Washington, D.C. 20402.

2. Copies may be obtained from: American National Standards Institute, 1430 Broadway, New York, New York 10018.

3. This is proportional two a gear development greater than 6.67 m (21:9 ft) In the bicycle's highest gear ratio. Gear Development is the distance the bicycle travels in meters in one crank revolution.

Appendix D

Engineering Bibliography

A Survey of Characteristics Associated With Emergency Medical Services-Reported Bicycle-Related Injuries in King County, Seattle-King County Department of Public Health, WA, October 1985, 35 pages.

Abbott, A. V., "Second International Human Powered Vehicle Scientific Symposium," INTERNATIONAL HUMAN POWERED VEHICLE ASSOCIATION, 1984, 204 Pages.

Adams, W. C., "Influence of Age, Sex, and Body Weight on the Energy Expenditure of Bicycle Riding," JOURNAL OF APPL. PHYSIOLOGY, Vol. 22(3) March 1967, Pages 539-545.

Adeyefa, B. A., "Determination of the Loads, Deflections and Stresses in Bicycle Frames," DISSERTATION, UNIVERSITY OF MANCHESTER, March 1978, 133 Pages.

Allen, C. E. and Borland, S. W., "Bicycle Safety," AMERICAN JOURNAL OF DISEASES OF CHILDREN, February 1987, Vol. 141, Pages 136-137.

Annotated Bibliography of Documents Available from BHSI Documentation Center, Bicycle Helmet Safety Institute DOC #168, Arlington, VA, 26 pages.

Archer, J.; Hamley, E. J., and Robson, H. E., "The Fitness Assessment of Competitive Cyclists," INTERNATIONAL CONGRESS ON THE PSYCHOLOGY OF SPORT, Pages 51-57.

Ashton, S. J., "Pedestrian Accident Investigation and Reconstruction," SEMINAR ON SPECIAL PROBLEMS IN TRAFFIC ACCIDENT RECONSTRUCTION, Institute of Police Technology and Management, University of North Florida, April 10-14, 1989, 37 Pages.

Ashton, S. J., and Lambourn, R. F., "Some Aspects of Two Wheeled Vehicle Accident Reconstruction," 1989 NATIONAL POLICE ACCIDENT INVESTIGATION SEMINAR, Hendon, U.K., March 16,17, 1989, 23 Pages.

Ashton, S.J., Pedder, J.B., and Mackay, G.M., Pedestrian Injuries and the Car Exterior, Dept. of Transportation and Environmental Planning, University of Birmingham (England), SAE Paper 770092, International Congress and Exposition, February 28 - March 4, 1977.

Austin, R. L.; Klassen, D. J., and Vastrum, R. C.., "Pedestrian Conspicuity Under the Standard Headlight System Related to Driver Perception," THE TRANSPORTATION RESEARCH BOARD, JANUARY 1975, Pages 1-23, A1-A4.

Automotive Frontal Impacts SP-782, SOCIETY OF AUTOMOTIVE ENGINEERS, INC., February 1989, 144 Pages.

Beier, G.; Schuck, M.; Schuller, E., and Spann, W., "Determination of Physical Data of the Head I. Center of Gravity and Moments of Inertia of Human Heads," INSTITUTE OF FORENSIC MEDICINE, Univ. of Munich, WG, April 1, 1979, Pages 1-35.

Berger, R. E., "Considerations in Developing Test Methods for Protective Headgear," US DEPT OF COMMERCE, PRODUCT SAFETY ENGINEERING SECTION, NBSIR 76-1107, August 1976, 53 Pages.

Bicycle Fork End Construction, U. S. Patent # 4,121,850, October 24, 1978, Dept. of Commerce, U.S. Patent and Trademark Office, Inventor, Sherwood B. Ross, 5 pages.

Bicycle-Related Portions of the Uniform Vehicle Code, with 1984 revisions, Bicycle USA, Baltimore, MD, 20 pages.

Bicycle-Safe Grate Inlets Study, Volume 1 Hydraulic and Safety Characteristics of Selected Grate Inlets on Continuous Grades, US DEPT OF COMMERCE, NATIONAL TECHNICAL INFORMATION SERVICE, June 1977, 288 Pages.

Bicycle/Pedestrian Planning & Design, AMERICAN SOCIETY OF CIVIL ENGINEERS, 1974, 700 Pages.

Bishop, P. and Briard, B., "Helmets Put to the Test," BICYCLING NEWS CANADA, Spring 1984, Pages 12-13.

Blomberg, R. D.; Hale, A., and Preusser, D. F., "Conspicuity for Pedestrians and Bicyclists: Definition of the Problem, Development and Test of Countermeasures," US DEPT. OF TRANSPORTATION, NATIONAL HWY TRAFFFIC SAFETY ADM., DOT HS-806 563, April 1984, Pages i-xii, l-78, Appendix A-l - K-5.

Blomberg, R. D.; Hale, A., and Preusser, D. F., "Experimental Evaluation of Alternative Conspicuity-Enhancement Techniques for Pedestrians and Bicyclists," JOURNAL OF SAFETY RESEARCH, 1986, Vol. 17, Pages 1-12.

Blomberg, R. D.; Leaf, W. A., and Jacobs, H. H., "Detection and Recognition of Pedestrians at Night," DUNLAP AND ASSOCIATES, INC., June 24, 1980, Pages 17-21.

Bloom, S. R.; Johnson, R. H.; Park, D, M.; Rennie, M. J., and Sulaiman, W. R., "Differences in the Metabolic and Hormonal Response to Exercise Between Racing Cyclists and Untrained Individuals," J. PHYSIOLOGY, 1976, 258, Pages 1-18.

BMX bike injuries: the latest epidemic, British Medical Journal, Volume 289, October 13, 1984, Pages 960-96 1.

Burgett, A. and Villalba, V., "A Short Study of Seeing Distance," SOCIETY OF AUTOMOTIVE ENGINEERS, INC., 1985, 14 Pages.

Burke, Ed, Ph.D., and Fritschner, Diane, USCF Director, "Get A Head Start - Head Injury, Helmets and the Cyclist," USCF, Colorado Springs, CO, November 1987, 11 pages.

"Causative Factors & Countermeasures for Rural and Suburban Pedestrian Accidents Accident Data Collection and Analysis," US DEPT OF COMMERCE, NATIONAL TECH INFORMATION SERVICE, March 1977, 266 Pages.

Chapman, R.G., "Accidents on urban arterial roads", Transport and Road Research Laboratory Report 838,, Crowthorne, Berkshire, 1978, 11 pages.

Child Bicycle-Mounted Seats: the Risks and the Rules, PRO BIKE NEWS, September 1988, Vol. 8(9) 1 Page.

Craig, A. B. Jr., "Effects of Position on Expiratory Reserve Volume of the Lungs," JOURNAL APPL. PHYSIOLOGY, January 1960, Volume 15(l), Pages 59-61

Cross, K., "Bike Ed '77," US DEPT OF TRANSPORTATION AND US CONSUMER PRODUCT SAFETY COMMISSION, May 4-6, 1977, Pages 34-49.

Cycling as a mode of transport, Transport and Road Research Laboratory Symposium Supplementary Report 540, 1978, Crowthorne, Berkshire, 1980, 108 pages.

Dal Monte, A.; Manoni, A., and Fucci, S., "Biomechanical Study of Competitive Cycling," MEDICINE AND SPORT, BIOMECHANICS III, Karger, Basel 1973, Vol. 8, Pages 434-439.

Daly, D. J. and Cavanagh, P. R., "Asymmetry in Bicycle Ergometer Pedalling," MEDICINE AND SCIENCE IN SPORTS, 1976, Vol. 8(3), Pages 204-208.

DI Prampero, P. E.; Cortili, G.; Mognoni, P., and Saibene F. "Equation of Motion of a Cyclist," JOURNAL OF APPLIED PHYSIOLOGY, July 1979, Vol. 47(l), Pages 201-206.

Dixon, H. S., "Nighttime Automobile/Pedestrian Accidents; Driver View: Whence Pedestrian?" NIGHTTIME AUTOMOBILE NAFE 067D, December 1987, Pages 63-71.

Ewing, C. L. and Thomas D. J., "Human Head and Neck Response to Impact Acceleration," ARMY - NAVY JOINT REPORT, BICYCLE HELMET SAFETY INSTITUTE BHSIDOC #249, August 1972, Pages 1-215.

Faria, I. E., "Cycling Physiology for the Serious Cyclist," CHARLES C. THOMAS, BANNERSTONE HOUSE, 1978, 162 Pages.

Fife, D.; Davis, J.; Tate, L.; Wells, J. K.; Mohan, D., and Williams, A., "Fatal Injuries to Bicyclists: The Experience of Dade County, Florida," THE JOURNAL OF TRAUMA, August 1983, Vol. 23(8), Pages 745-755.

Fleming, A., "Bicycles", INSURANCE INSTITUTE FOR HIGHWAY SAFETY, July 1989.

Flora, J. D. and Abbott, R.D., "National Trends in Bicycle Accidents," JOURNAL OF SAFETY RESEARCH, Spring 1979, Vol. 1 l(l), Pages 20-27.

Fontanet, P., Vintre, P., and Trundle, "Moyens d'essais utilises par la mise au point des vehicules, complets," S.I.A., 5 May 1974.

Forester, J., "Bicycle Transportation," THE MIT PRESS, Cambridge, Mass., London, Eng.

Forester, J., "Forensic Engineering in Bicycle Accidents," Pub. 3 10, Chapter 5 0 ,Pages 50- 1-50-82.

Fricke, Lynn B., "Vehicle-Pedestrian Accident Reconstruction." ACCIDENT RECONSTRUCTION JOURNAL, January/February 1992, Pages 18-29.

Garriott, James C., MEDICOLEGAL ASPECTS OF ALCOHOL DETERMINATION IN BIOLOGICAL SPECIMENS, Lawyers and Judges Publishing Company, 1993.

Get Smart Head Shop (The), or How To Choose A Helmet Intelligently, BICYCLING NEWS CANADA, Spring 1984, Pages 16-17.

Gibson-Harris, Sheree, P.E., "Would a Warning Have Prevented the Accident?" NAFE 185A, Pages 51-55.

Gisolfi, C. V.; Rohlf, D. P.; Navarude, S. N.; Hayes, C. L., and Sayeed, S. A., "Effects of Wearing a Helmet on Thermal Balance While Cycling in the Heat," THE PHYSICIAN AND SPORTSMEDICINE, January 1988, Vol. 16(l), Pages 139-146.

Godthelp, J. and Wouters, P. I. J., "Course Holding by Cyclists and Moped Riders," APPLIED ERGONOMICS, December 1980, 11.4, Pages 227-235.

Gollnick, P. D.; Piehl K., and Saltin, B., "Selective Glycogen Depletion Pattern in Human Muscle Fibres After Exercise of Varying Intensity and at Varying Pedalling Rates," J. PHYSIOLOGY, 1974, 241, Pages 45-57.

Gueli, D. and Shephard, R. J., "Pedal Frequency in Bicycle Ergometry," CANADIAN JOURNAL OF APPLIED SPORT SCIENCES, 1976, Vol. 1, Pages 137-141.

Guide for Development of New Bicycle Facilities 1981, AMERICAN ASSOCIATION OF STATE HIGHWAY AND TRANSPORTATION OFFICIALS, October 3, 1981, Pages 1-3 1.

Gurdjian, E. S.; Roberts, V. L., and Thomas, L. M., "Tolerance Curves of Acceleration and Intracranial Pressure and Protective Index in Experimental Head Injury," THE JOURNAL OF TRAUMA, 1966, Vol.8(5), Pages 600-604.

Hagberg, J. M.; Mullin, J. P.; Giese, M.D., and Spitznagel, E., "Effect of Pedaling Rate on Submaximal Exercise Responses of Competitive Cyclists," JOURNAL OF APPLIED PHYSIOLOGY, 1981, Vol. 51(2), Pages 447-451.

Haight, W. R., and Eubanks, J. J., "Trajectory Analysis for Collisions Involving Bicycles and Automobiles," INTERNATIONAL CONGRESS AND EXPOSITION, Detroit, Michigan, March 2, 1990, Pages 143-161.

Hale, A. and Zeidler, P., "Review of the Literature and Programs for Pedestrian and Bicyclist Conspicuity," DUNLAP AND ASSOCIATES EAST, INC., April 1984, Pages 1-73. Hard vs. Soft, Velo-News, June 11, 1990, 1 page.

Hanson, B.D. "Wet-Weather-Effective Bicycle Rim Brake: An Exercise in Product Development," M.S. THESIS, Massachussets Institute of Technology, 197 1.

Harger, R.N., Hulpieu, R.H., and Lamb, E.B., "The Speed With Which Various Parts of the Body Reach Equilibrium in the Storage of Ethyl Alcohol." J. Biol. Chem. 120:689-704, 1937.

Hartung, H. and McMillen, J., "Specificity of Aerobic Testing in Competitive Cyclists Compared With Runners," EXERCISE PHYSIOLOGY, Pages 615-622.

Hayduk, D., "Bicycle Metallurgy for the Cyclist," JOHNSON PUBLISHING CO., Boulder, July 1987, 111 Pages.

Hazard Analysis - Bicycles, July 1973, U.S. Consumer Product Safety Commission, Bureau of Epidemiology, Washington, D. C. 20207 Helmet Adhesion Study Shows Pros, Cons for Both Types, Bicycle Dealer Showcase, April 1990, 1 page.

Henderson, G. R., "Calibration Methodology for Protective Headgear Drop Test Equipment," BICYCLE HELMET SAFETY INSTITUTE BHSIDOC #24, May 10, 1975, Pages 1-19.

Higgins, Mark; Kick, James; Plant, David; Ries, Douglas; and Weiss, Kathleen, "Determination of Maximum Vehicle Speed in Pedestrian Impacts," ACCIDENT RECONSTRUCTION JOURNAL, July/August 1990, page 18.

High-Rise Bicycles, National Commission on Product Safety Final Report Presented to the President and Congress June 30, 1970.

Hill, P, F., "Bicycle Law and Practice," BICYCLE LAW BOOKS, 1986, 238 Pages.

Hjertberg, Eric, "The Wheel Scene - Mechanics Column, Cycling USA, March 1988, 1 page.

Hodges, Mark, "Bike Fit," Triathlete, July 1986, 8 pages.

Hoes, M. J. A. J. M.; Binkhorst, R. A.; Smeekes-Kuyl, A. E. M. C., and Vissers, A. C. A., "Measurement of Forces Exerted on Pedal and Crank During Work on a Bicycle Ergometer at Different Loads," INT. 2. ANGEW. PHYSIOL. EINSCHL. ARBEITSPHYSIOL., 1968, Vol.26, Pages 33-42.

Hoey, W. III; Kaplan, J.; Kraft, W.; Rogers, R., and Kronski, J., "Bicycle Transportation, A Civil Engineer's Notebook for Bicycle Facilities," AMERICAN SOCIETY OF CIVIL ENGINEERS, 1980, 189 Pages.

Hub Anti-Escape Device for a Bicycle, Invented by Takashi Segawa, Patented by Shimano Industrial Company, Ltd., U. S. Patent No. 4,079,958, March 1978, 8 pages.

Illingworth, Cynthia M, Dilys Noble, Deborah Bell, Ian Kemn, Ciaron Roche, Jennifer Pascoe, "150 bicycle injuries in children: a comparison with accidents due to other causes,"The British Journal of Accident Surgery, Volume 13, Number 1, Page 7-9.

Illingworth, Cynthia M., "Injuries to children riding BMX bikes," The British Medical Journal, Volume 289, October 13, 1984, Pages 956-957.

Joksch, H. C., "The Impact of Severe Penalties on Drinking and Driving," AAA FOUNDATION FOR TRAFFIC SAFETY, May 1988, 28 Pages.

Jordan, L. and Merrill, E. G., "Relative Efficiency as a Function of Pedalling Rate for Racing Cyclists," JOURNAL OF PHYSIOLOGY, 1979, Vol.296, Pages 49P-50P.

Juhl, M., "Bicycle Spoke Injuries," DEPARTMENT OF ORTHOPAEDIC SURGERY, Odense University Hospital, Odense, Denmark, September 7,8, 1976, Pages 23-32.

Kearney, Ed, "Rules For Bicycling At Night," Revision of rules published in the September 1976 issue of the L.A.W. Bulletin.

Kearney, Ed., "See and Be Seen with Real Bicycle Lights, May 1985, Washington, D. C., 4 pages.

Kell, S. O., "Bicycle-Mounted Child Seats are Risky Rides for Kids," FOR KIDS' SAKE, Winter 1989, Vol.6(4), 1 Page.

Li, Guohoa and S.P. Baker, "Alcohol in Fatally Injured Bicyclists," ACCID. ANAL. AND PREV., Vol. 26, No. 4, pp. 543-548, 1994.

Limpart, Rudolf, "Car-Pedestrian Accidents," ACCIDENT RECONSTRUCTION JOURNAL, January/February 1992, pages 18-29.

Mackie, A. M., "Research for a road safety campaign- accident studies for advertising formulation," Transport and Road Research Laboratory Report LR 432, Crowthome, Berkshire, 1972, 19 pages.

Mathematical Modeling Biodynamic Response to Impact SP-412, SOCIETY OF AUTOMOTIVE ENGINEERS, INC., October 1976, 96 Pages.

McHenry, S. R. and Wallace, M. J., "Evaluation of Wide Curb Lanes as Shared Lane Bicycle Facilities," MARYLAND STATE HIGHWAY ADMINISTRATION, August 1985, 90 Pages.

Merrill, E. G. and White, J. A., "Physiological Efficiency of Constant Power Output at Varying Pedal Rates, JOURNAL OF SPORTS SCIENCES, 1924,2, Pages 25-34.

Moll, Richard A., Ph.D.,P.E., "What Forensic Engineers Should Know about Products Liability Law," NAFE Journal, Page 57.

Montgomery, R. K., "Helmets Voice of Dissent," BICYCLING NEWS CANADA, Spring 1984, Pages 14-15.

Nicholas, M., "'Sate-Lite No.709' Spoke Mount-Bicycle Reflex Reflectors," ELECTRICAL TESTING LABORATORIES, INC. REPORT NO. 430713, February 3, 1975, Pages 1-12.

Nordeen-Snyder, K. S., "The Effect of Bicycle Seat Height Variation Upon Oxygen Consumption and Lower Limb Kinematics," MEDICINE AND SCIENCE IN SPORTS, 1977, Vol. 9(2), Pages 113-117.

Optical Performance of High-Visibility Garments and Accessories For Use on the Highway, BRITISH STANDARDS INSTITUTION, BS 6629, 1985, Pages 1-12.

Otto, D., "Injury Mechanism and Crash Kinematic of Cyclists in Accidents-An Analysis of Real Accidents," ACCIDENT RESEARCH UNIT, Medical University of Hannaver, Federal German Republic

Overstreet, J. T. Jr., "Baltimore County Bicycle-Moving Vehicle Accident Report for 1987," BALTIMORE BICYCLING CLUB, February 1989, Vol. 3, 34 Pages.

Overstreet, J. T. Jr., "The 1987 Maryland Bicycle Accident Report Less Baltimore City," BALTIMORE BICYCLING CLUB, July 1988, 22 Pages.

Overstreet, J. T. Jr., et al, "The Maryland Night Bicycle Accident Report for 1987,"

BALTIMORE BICYCLING CLUB, March/April, 1989, Vol. 1, 62 Pages.

Pandolf, K. B., and Noble, B.J., "The Effect of Pedalling Speed and Resistance Changes on Perceived Exertion for Equivalent Power Outputs on the Bicycle Ergometer," MEDICINE AND SCIENCE IN SPORTS, 1973, Vol. 5 No. 2, Pages 132-136.

Papadopoulos, Jim M., "Bicycle Steering Dynamics and Self-Stability: A Summary Report On Work In Progress, Preliminary Draft, December 1987, 41 Pgs.

Peters, George A., "Warning Signs and Safety Instructions: Covering all the Bases, Risk Management, September 1985, 4 pages.

Peters, George A., Esq., P.E., "Warnings and Instructions For The Packaging Specialist," Hazard Prevention, January/February 1986, 2 pages.

Peters, George A., Peters, Barbara J., "Automotive Engineering and Litigation," Garland Law Publishing, NY, 1986, 4 pages.

Protect Young Cyclists, FAMILY SAFETY & HEALTH, Fall 1987, Page 31. Quick-Release Hub Retention Device, U.S. Patent 4,103,922, U.S. Department of Commerce, U.S. Patent and Trademark Office, August 1, 1978, Inventor Frank P. Brilando, 8 pages.

Report on 1981 Pedestrian and Bicycle Accidents in Washington, D.C. TRANSPORTATION SAFETY DIV., DISTRICT OF COLUMBIA's DEPARTMENT OF TRANSPORTATION, January 1982, Pages 1, 17-29.

Revised Safety Standards - Bicycles, Consumer Product Safety Commission, December 1978 Part IV, A-13 -1-23.

Revised Safety Standards - Bicycles, Consumer Product Safety Commission, Part IV Federal Register, December 22, 1978.

Rhoades, K., "Bicycle Carrier Seats," PROTECT YOURSELF, May 1987, Pages 12-17.

Rice, R. S. and Roland, R. D. Jr., "An Evaluation of the Performance and Handling Qualities of Bicycles," CORNELL AERONAUTICAL LABORATORY, INC., Cal No. VJ-2888-K, April 1970, 67 Pages.

Road Accidents Great Britain 1980, Department of Transport, Scottish Development Department, Welsh Office Ruderman, S., "Bicycle Injuries in a Suburban Community," BICYCLE HELMET SAFETY INSTITUTE BHSIDOC #42, August 3, 1987, Pages l-9.

SAE Ground Vehicle Lighting Manual (1986), SOCIETY OF AUTOMOTIVE ENGINEERS, INC., April 1986, 266 Pages.

SAE Ground Vehicle Lighting Manual (1987), SOCIETY OF AUTOMOTIVE ENGINEERS, INC., April 1987, 294 Pages.

SAE Handbook(1986), INDEX," SOCIETY OF AUTOMOTIVE ENGINEERS, 1986,187 Pages.

Safety Mounting for Quick-Release Hubs, U.S. Patent 3,807,761, U.S. Department of Commerce, U.S. Patent and Trademark Office, April 30, 1974, Inventors: Frank P. Brilando, Stanley R. Jameson Sargent, J. D.; Peck, M.G., and Weitzman, M., "Bicycle-Mounted Child Seats," AMERICAN JOURNAL OF DISEASES OF CHILDREN, July 1988, Vol. 142, Pages 765-767.

Schubert, J., "Should You Be Wearing Early Warning?" BICYCLING, Pages 52-56.

Schmidt, David H. and Nagal, Donald A., "Pedestrian Impact Case Study," Proceedings of the Fifteenth Conference of the American Association for Automotive Medicine, October 20-23, 1971, Pages 151-167.

Seabury, J. J.; Adams, W. C., and Ramey, M. R., "Influence of Pedalling Rate and Power Output on Energy Expenditure During Bicycle Ergometry," ERGONOMICS, 1977, Vol. 20(5), Pages 491-498.

Searle, John A. and Searle, Angela, "The Trajectories of Pedestrians, Motorcycles, Motorcyclists, etc, Following a Road Accident," (1983) SAE PATER NO. 831622.

Selbst, S. M.; Alexander, D., and Ruddy, R., "Bicycle-Related Injuries." AMERICAN JOURNAL FOR DISEASES OF CHILDREN-VOL 141, February 1987, Pages 140-144.

Sellers, D., "Reflectors and Reflective Materials," BIKE TECH, Pages 7-13.

Sellers, David, "Believing is Seeing - Bicycle Headlights Test Report, Bike Tech, Winter 1985, Vol. 4, No. 5, 8 pages.

Sellers, David, "Reflectors and Reflective Materials," Bike Tech, June 1985, Vol. 4, Number 3, 7 pages.

Sharp, A., "Bicycles & Tricycles, An Elementary Treatise on Their Design and Construction," THE MIT PRESS, Cambridge, Mass and London, Eng., 1984, 536 Pages.

Small Parts Gage - Requirements for Bicycles, Consumer Product Safety Commission, 16 CFR Ch.11 (1-1-87 Edition), Pages 442-465

Soden, P. D. and Adeyefa, B. A., "Forces Applied to a Bicycle During Normal Cycling," J. BIOMECHANICS, GB 1979, Vol. 12, Pages 527-541.

Sparnon, A.L., Ford, W.D.A., "Bicycle Handlebar Injuries in Children," Journal of Pediatric Surgery, Vol21, No 2, February 1986, Pages 118119.

Sparon, Tony, Moretti, Kim, Sach, Randall P., "BMX handlebar - A threat to manhood?" The Medical Journal of Australia, December 1982, 2 pages.

Specification for Photometric and Physical Requirements of Reflective Devices Part 2. BRITISH \STANDARD CYCLES, BS 6102, 1982, Pages 1-6.

Stutts, J.C.; Williamson, J. E., and Sheldon, F. C., "Bicycle Accidents: An Examination of Hospital Emergency Room Reports and Comparison with Police Accident Data," UNC HWY SAFETY RESEARCH CENTER, January 1988, 32 Pages.

Tarriere, C., Fayon, A., Hartmann, F. and Ventre, P., "The Contribution of Physical Analysis of Accidents Towards Interpretation of Severe Traffic Trauma," STAPP, San Diego, California, 17-19 November 1975.

Thompson, R. S.; Rivara, F. P., and Thompson, D. C., "A Case-Control Study of the Effectiveness of Bicycle Safety Helmets," THE NEW ENGLAND JOURNAL OF MEDICINE, May 25, 1989, Vol. 320(21), Pages 1361-1367.

Ulmer, R. G.; Leaf, W. A., and Blomberg, R. D., "Analysis of the Dismounted Motorist and Road Worker Model Pedestrian Safety Regulations, DUNLAP AND ASSOCIATES, INC., August 1982, Pages i-xv, 1-63, A-1 - C-5.

Van der Plas, R., "Staying Alive at Night, An Evaluation of Bicycle Lighting Systems," BICYCLING, DECEMBER 1978, Pages 49-53,62-65.

Van der Plas, R., "Staying Alive at Night, More About Bike Lights," BICYCLING, January 1979, Pages 53-59.

Ventre, P. and Provensal, J., "Proposition d'une methode d'analyse et de classification des severites de collisions en accidents reels," IRCOBI, Amsterdam 26-27 June 1973.

Viscous Criterion (The), C&EN, Wash.,DC. January 9, 1989.

von Dobeln, W., "A Simple Bicycle Ergometer," JOURNAL OF APPLIED PHYSIOLOGY, September 1954, Vol. 7(2), Pages 222-224.

Wachtel, A., "Bicycle Helmet Safety," INSTITUTE BHSIDOC #130, March 1982, Pages 1-7.

Watts, G. R., "Pedal Cycle Lamps and Reflectors-Some Visibility Tests and Surveys," TRANSPORT AND ROAD RESEARCH LABORATORY REPORT 1108, 1984, Pages 1-19.

Watts, G.R., "Bicycle safety devices-effects on vehicle passing distances," Transport and Road Research Laboratory Supplementary Report 512, Crowthome, Berkshire, 1979, 6 pages.

Watts, G.R., "Evaluation of pedal cycle spacers," Transport and Road Research Laboratory Supplementary Report 820, Crowthrone, Berkshire, 1984, 8 pages.

Weiss, B. D., "Childhood Bicycle Injuries - What Can We Do?" AMERICAN JOURNAL OF DISEASES OF CHILDREN, February 1987, Vol. 141, Pages 135-136.

Whitt, Frank Rowland and Wilson, David Gordon, "Bicycling Science," MIT, 1982.

Williams, A. F., "Factors in the Initiation of Bicycle-Motor Vehicle Collisions," AMERICAN JOURNAL OF DISEASES OF CHILDREN, April 1976, Pages 370-377.

Woodson, Wesley E., HUMAN FACTORS DESIGN HANDBOOK, Second Edition, McGraw Hill, 1992, Pages 541-542.

Wright, R., "Building Bicycle Wheels, Step-By-Step Instructions for Building, Repairing, and Maintaining Spokes, Rims, and Hubs," MACMILLIAN PUBLISHING COMPANY, 1977, 46 Pages.

Zahradnik, Fred and Redcay, Jim, "Here's the Answer to Which of 18 Popular Brakes Work Best," Bicycling, February 1987, Pages 70-73.

About the Authors

Mr. James M. Green, P.E., DEE, is President of GE Engineering, Inc., in Asheville, North Carolina, a general Civil Engineering firm with a specialty in bicycle accident reconstruction. He is a member of the National Society of Professional Engineers and a Fellow in the American Society of Civil Engineers and the National Academy of Forensic Engineers (NAFE). He is a registered Professional Engineer in multiple states. The author is an active racer on the triathlon circuit and has qualified for and completed the Hawaiian Ironman Triathlon three times. He was two-time U.S. Cycling Federation District Road Champion, as well as the State of Tennessee Triathlon Champion. Mr. Green, a bicycle builder and tester, supplements his road bicycle training with off-road bicycle training. He has investigated hundreds of bicycle accidents during his more than twenty years of professional practice.

Contact information:
    120 Kalmia Dr.
    Asheville, NC 28804
    Phone: (828) 236-1492
    Fax: (828) 236-0355
    E-mail: setg@mindspring.com
    Web site: www.bicyclereconstruction.com

Robert Mionski, Esq. is an attorney licensed to practice law in Minnesota, Oregon, and Wisconsin. His practice consists primarily of bicycle accident litigation. He also represents professional cyclists, who are members of such professional teams as U.S. Postal Service, Telekom, Mercury, Saturn, Prime Alliance, and JellyBelly. A former U.S. Olympic Cycling Team member and National Road Champion, Mr. Mionske still enjoys riding his bicycle several times a week.

Anand David Kasbekar, Ph.D., is a consulting engineer for Research Engineers, Inc. and President of Visual Sciences, Inc. in Raleigh, North Carolina. As a consulting engineer at Research Engineers, Inc. (REI) since 1987, Dr. Kasbekar has expertise in the areas of forensic engineering, materials characterization, product liability, and failure analysis. He developed the visualization laboratory for REI and has over eight years of experience in the application of computer simulation and visualization to the fields of engineering and science. Additionally, Dr. Kasbekar serves as an Adjunct Assistant Professor in the Department of Mechanical Engineering and Materials Science at The Duke University School of Engineering in Durham, North Carolina. He is an active member in several engineering societies including the Society of Automotive Engineers (SAE), The National Safety Council (NSC), The Association for Computing Machinery (ACM), and the American Society of Materials (ASM).

Kraig Willett is a Washington State native currently residing in the greater San Diego area. Better known in cycling circles as "Kirk's Brother", he is a Product Development Engineer in the golf industry and has previously worked as an engineer in the bicycle industry. He graduated Summa Cum Laude with a B.S. in Mechanical Engineering and a Minor in Materials Science and Engineering from Washington State University. After a failed attempt to become a pro bike racer, he spent two years at Virginia Tech working as a research assistant while pursuing a Master's degree in Aerospace Engineering. He is a former Category 1 cyclist on the road and currently commutes by bicycle to work every day. Engineering is his profession, but bikes and the people who ride them are his passion.

INDEX

A

acceleration- 28
   constant- 90
acceleration forces- 17
accident- 1, 5, 7, 12, 17, 18, 19, 20, 21, 24, 27, 31, 44, 56, 60, 63, 66
   nighttime- 213
Aerodynamic drag- 257
aerodynamics- 261
alcohol- 213, 214, 215, 216
all terrain bicycles (ATB)- 110
aluminum- 6, 7, 230
ambient- 81, 241
ambient conditions- 62
ambient lighting- 55
American Railway Engineering Association (AREA)- 52
analysis
   photometric- 77
   trajectory- 76
angle of impact- 229
ANSI- 17, 18, 19, 21
Appel- 147
ASTM- 18, 19, 21, 60, 61
ASTM Standards For Cycle Helmets- 17
average force- 165
avoidance- 39, 78, 82

B

Barnes v. City of Tucson- 345
baron carbide- 232
barricade- 220
Bennett v USCF- 220
Bennett v. USCF- 347
Bernoulli's Principal- 248
Bernoulli's Principle- 244, 246
biathlon- 219

bicycle- 213
   races- 219
   regulations- 382
   requirements- 381
bicycle accident- 11, 23, 51, 55
   daytime- 66, 67
   nighttime- 66, 67, 73
bicycle helmet- 17, 18
bicycle mechanic- 5
bike- 5, 42, 51
   mountain- 234
bike fit- 251
bike frame size- 252
bike path- 23, 24
Bikeways
   Class I- 23
biodynamics
   of the fall- 43
biokinetics- 20, 21, 41
Bledsoe v. Goodfarb & Salt River Valley Water Users Assoc.- 348
Blomberg- 59, 65, 66, 82, 83, 239, 240
Blomberg, Richard D.- 58
blood alcohol content- 213
blunt trauma- 17
brakes
   caliper- 103
braking
   distance- 81, 104, 108
braking system
   requirements- 384
Brilando device- 43
Brilando patent- 52
Brown and Obenski- 67
Buchan V. USCF- 350
Burns' and Sullivan- 92
buses- 243

C

Campagnolo- 7, 13
Campagnolo Record QR- 197
Cannondale- 3, 4, 230
Caporale V. Raleigh Industries of America- 351
car-bicycle accidents- 31, 65, 67
causal factors- 65
center of gravity- 104, 123, 128, 129, 204, 224
center of mass- 100
chain
    to jump- 14
chain stays- 7
chrome-moly- 124
Civil Engineering standards- 25, 52
civil liability- 189
clothing
    fluorescent- 65
Code of Federal Regulations- 381
Coefficient of Drag- 91
coefficient of friction- 54, 104, 133, 146
collapse- 16, 112
    of the front wheel- 8
collision- 31
    nighttime- 63
color- 59
component- 99, 101
    failures- 229
component failure- 5, 6, 16
composite- 7
composite materials- 232, 233
conspicuity- 1, 31, 55, 58, 59, 63, 65, 66, 67, 73, 172, 237, 241
    night-time- 62
Constuction guidleines
    highway-railway crossings- 333
Consumer Product Safety Commission- 16, 60, 78, 390
    policies & interpretations- 390

Consumer Product Safety Commission (CPSC)- 83
contrast- 74
crush data- 171, 175
cycle racers- 54
cyclist- 21, 28, 31, 32, 39, 41, 51, 52, 57, 64, 124, 128, 213
Cyclists Overtaken by Motorists- 267

D

deceleration- 27, 28, 104, 105, 107
deformation- 165
Dennis v. City of Tampa- 353
Department of Transportation- 77, 83
derailleur- 12
design speed- 23
drainage culvert- 112
drive chain
    requirements- 388
dropouts- 7

E

EDSMAC- 2, 203
EDSVS- 31
electromagnetic field- 56
electromagnetic wave- 56
Entering or Crossing Roadways- 268
equation- 88, 91, 95, 96, 97, 100, 102, 107, 123, 124, 125, 134

F

failure- 42, 110
    of the seat post- 12
    pressure- 111
fall- 21
    to the side- 223
fatigue points- 6
federal conspicuity requirements- 60
fiction shifter- 13
fit- 5
fluorescent- 66

force
    static- 112
force of impact- 2, 27, 29, 167
force of landing- 20
fork deflection- 167, 233
frame- 6, 43, 109, 111, 124, 167
    failures- 229
frame alignment tool- 7
frame failure- 6, 7
Frame Measurement- 251
frames- 7
frequency- 59
friction- 89
front fender- 11
front fork- 41, 44, 51, 111, 128
    closed end- 43
    deflection- 166, 175
    deformation- 171
front wheel- 128
    disengaging- 51
    failure of- 8
frontal forces- 233
frontal impact- 167

G
Garriott- 216
    James- 214
Goshen Luna Pro- 61
Gravitational forces- 257
gridding the site- 75, 76
Griffen USA- 232

H
Haight and Eubanks- 151
handling characteristics- 229
Haverstock v. Raleigh- 354
head- 20
head trauma- 21
headlight- 62
helmet- 1, 17
    colors- 18
    hard shell- 19
    soft-shell- 19
Hiett v Lake Barcroft Community Association- 221
Hiett v. Lake Barcroft Community Assoc.- 355
Hillis v. City of Oneonta- 357
Hozan tension meter- 9

I
illumination- 75, 76, 77
    analysis- 62
impact- 141
    force- 134
    post- 203
    speed- 167, 171, 175, 233
    velocity- 133
index shifter- 14
inertia- 87
Inertial forces- 257
injuries- 17
injury- 21
instantaneous maximum force- 165
instructions and labeling- 388
Intersection Accident- 268
intersection configurations- 39

J
Johnson- 57, 60, 171, 175
Johnson vs. Derby Cycle Corp.- 2
Johnson's Admix v. Richmond and Danville R.R. Co- 221
Johnston v. Lynch- 358

K
Kestrel- 232
kinematics- 133, 203
kinetic energy
    rotational- 90
    translational- 90
Klein- 230

L
Laird v. Kostman- 360
law of motion
    second- 257
Li- 216
Li study- 213, 214
light- 55, 57, 73
Litespeed- 229
Look- 195
loss of balance- 223, 225
lug- 6
luminance- 74, 77
lux- 73

M
Madison & YMCA of Metroploitan LA v. Superior Court of the State of California & Sulejmanagic- 361
manufacturer- 67, 109, 172
Mavic 646LMS- 196
McWhorter v. Clark- 364
mechanic- 7
metallurgist- 6, 7, 12
metallurgy- 229
Moore v. Swoboda- 366
motor vehicle speed- 144
motorist- 39

N
National & International Brotherhood of Street Racers v. Superior Court of LA County & Castro- 352
neck- 19
Newton- 257

O
Olson and Sivak- 238
Olson, Paul- 237

P
Panitz vs. Behrend and Emsberger- 189
Papadopoulos, Jim M- 43
pedal
    clipless- 196
    requirements- 387
photometer- 61, 73
point of impact- 75, 151, 159, 162, 203
police accident report- 97
positive retention device- 52
Power- 257
power requirements- 261
premature release- 1, 43
    of the front wheel- 41, 42, 52
    of the rear wheel- 42
protective guards
    requirements- 388

Q
Quick Release Hub Retention Device- 42
quick release mechanism- 41, 43

R
railroad crossing- 51, 52, 53, 54
Raleigh- 3, 4, 91
Ralston v. Plogger- 368
Reach Adjustment- 253
reaction time
    perceived- 237
reaction times- 32, 65
rear derailleur- 13, 14
recognition- 78
recognition distances
    nighttime- 240
reflector- 76, 78, 83
    CPSC- 65
Renault study- 171
retroreflector- 2, 60, 61, 62, 65
Reynolds 753- 229
roadway- 23

Rolling resistance- 257
rotation- 124
rotational motion- 134
rotational velocity- 133

S
S-1 Gard- 248
Schaffner v. Chicago & Northwestern- 370
Schmidt- 141
Schmidt method- 143, 144, 151
Schwinn- 3, 42, 52
Searle method- 145, 146, 147, 151
Seat Adjustment- 253
Sellers- 66
Sellers, David- 64
shear- 100, 101, 102
Shimano- 13, 14
Sidewalk, Crosswalk Riding- 268
site survey- 97
slope
    elevation- 97
Smith v. Bradford- 372
Snell- 17, 18, 19, 21
speed of impact- 136
Speed Play Bryne Xl2- 197
spinal injuries- 20
spoke tension- 9
SRAM- 4
standard- 19, 60
    AREA- 53, 54
    ASTM- 61
    CPSC- 63
static testing- 109
statutory visibility requirements- 58
steel- 231
steering system
    requirements- 387
Swanlund v. Shimano Industrial Corp., LTD- 374

T
Taiwanese- 110
testing
    crush- 175
titanium- 229
topography- 66
torque- 42
total reaction time- 237, 239, 241
trajectory- 32, 99, 100, 143, 233
    simulation- 205
translational motion- 134
triathletes- 54
triathlon- 219, 232
Triathlon Federation- 221

U
Uniform Motor Vehicle Code- 57, 63
United States Cycling Federation- 219

V
vectors
    position- 174
    speed- 174
Vehicle Door and Unsafe Loads- 268
velocity- 95, 99, 101, 128, 134, 175, 206
    impact- 171
    projection- 145
visibility- 58, 74, 81
    daytime- 240
    nighttime- 240

W
Walker v. City of Scottsdale- 376
Weight distribution- 251
weight distribution- 173
Wellgo RG9203- 196
wheel
    aerodynamic- 259
    inertia- 259
    mass- 259

rotating- 243
wheel collapse- 159
wheel performance- 258
wheel well guard- 246
Whitt and Wilson- 108
wind resistance- 91
Wrong-Way Cyclists- 268

Z
Zone of Containment- 82

Printed in the United States
By Bookmasters